# Springer Undergraduate Mathematics Series

**Editor-in-Chief**

Endre Süli, Oxford, UK

**Series Editors**

Mark A. J. Chaplain, St. Andrews, UK

Angus MacIntyre, Edinburgh, UK

Shahn Majid, London, UK

Nicole Snashall, Leicester, UK

Michael R. Tehranchi, Cambridge, UK

The Springer Undergraduate Mathematics Series (SUMS) is a series designed for undergraduates in mathematics and the sciences worldwide. From core foundational material to final year topics, SUMS books take a fresh and modern approach. Textual explanations are supported by a wealth of examples, problems and fully-worked solutions, with particular attention paid to universal areas of difficulty. These practical and concise texts are designed for a one- or two-semester course but the self-study approach makes them ideal for independent use.

Stephan Rosebrock

# Visual Group Theory

## A Computer-Oriented Geometric Introduction

 Springer

Stephan Rosebrock
Institut für Mathematik und Informatik
Pädagogische Hochschule Karlsruhe
Karlsruhe, Baden-Württemberg, Germany

ISSN 1615-2085             ISSN 2197-4144    (electronic)
Springer Undergraduate Mathematics Series
ISBN 978-3-662-69364-3          ISBN 978-3-662-69365-0    (eBook)
https://doi.org/10.1007/978-3-662-69365-0

Mathematics Subject Classification: 20-01, 20-08

Translation from the German language edition: "Anschauliche Gruppentheorie. Eine computerorientierte geometrische Einführung" by Stephan Rosebrock. © Springer-Verlag GmbH Deutschland, ein Teil von Springer Nature 2020.
The original submitted manuscript has been translated into English. The translation was done using artificial intelligence. A subsequent revision was performed by the author(s) to further refine the work and to ensure that the translation is appropriate concerning content and scientific correctness. It may, however, read stylistically different from a conventional translation.
Translation from the German language edition: "Anschauliche Gruppentheorie" by Stephan Rosebrock, © Springer-Verlag GmbH Deutschland, ein Teil von Springer Nature 2020. Published by Springer Berlin Heidelberg. All Rights Reserved.

This Springer imprint is published by the registered company Springer-Verlag GmbH, DE, part of Springer Nature.
The registered company address is: Heidelberger Platz 3, 14197 Berlin, Germany

If disposing of this product, please recycle the paper.

*Julia gewidmet*

# Preface

*Numbers measure size, groups measure symmetry.*

*M. A. Armstrong*

Symmetry is everywhere. Flowers, crystals, many natural phenomena and artifacts appear beautiful to us, especially in their symmetrical appearance. One may enjoy this beauty and leave it at that in admiration. Here, a systematic approach to symmetry is proposed. Mathematically, it should be made as widely and as generally understandable as possible.

In the narrower sense, this book is about group theory. Groups can be considered as algebraic objects that describe the symmetry of geometric objects. This will mainly be the perspective adopted in the following, and thus, this book is also a book about geometry. Groups describe symmetry phenomena algebraically—one calculates with reflections, rotations, etc., generally with mappings of spaces onto themselves.

In group theory, definitions and theorems can often be formulated very concisely. The structure of group theory is clear and logical, and this is how it was usually taught not too long ago: many definitions and theorems supplemented with examples. However, this abstract presentation of group theory is neither genetically founded, nor is it structurally the simplest. The learning of new mathematical material, whether by children or adults, must build on prior knowledge—learning, as is well known, means integrating new knowledge into what we already know in a meaningful way, or reorganizing knowledge structures when the learning objects require it. For group theory, geometry serves as an excellent field of intuition. Reflections and sometimes rotations are familiar, they are already taught in elementary school and every student is well-acquainted with them. On the other hand, many books take their examples from the field of matrix groups. Matrices are typically encountered in the first two semesters of a mathematics degree, but often the concepts of a matrix and a linear mapping are not yet so firmly established that matrix groups can become illustrative examples of groups. Matrix groups are therefore only found in the appendix of this volume. Only there is knowledge of linear algebra assumed.

In addition, there are two further difficulties for the beginner in group theory: In school mathematics lessons, students are only confronted with abelian operations—even though, in my experience, one can start group theory meaningfully and successfully in the 5th grade of secondary school and in elementary school. So, beginner students are usually unfamiliar with non-abelian thinking due to certain deficiencies in their school career, and find it conceptually difficult. In addition, in school and at the beginning of our mathematical studies, we work with concrete objects (numbers, functions, matrices,...), and learners usually have to calculate something. In group theory, on the other hand, one proceeds axiomatically: "Imagine a set that fulfills this or that property". This approach places higher demands on the student's ability to abstract these ideas into mathematical thinking.

Now, however, geometry in group theory not only has high didactic value; modern developments in group theory also point in this direction. The theorem of Švarc–Milnor, the hyperbolic groups of Gromov, etc. interpret groups through their operations on geometric spaces; even more: The groups themselves become (via their Cayley graphs) geometric objects. This is a fundamentally different view of groups, which has been slowly gaining ground over the last 15 years and in turn builds on the work of Max Dehn and others from the beginning of the twentieth century. The basic ideas can already be found in the so-called "*Erlanger Programm*" of Felix Klein from the year 1872, in which he proposes to use groups for the classification of geometries (see Sect. 9.1).

These developments and fundamental reflections on mathematical didactics have inspired me to write the present volume. Firstly, an easily understandable introduction for all those who want to learn group theory should emerge. It therefore contains the essentials of the subject, without assuming much more than elementary secondary school level geometry and the basics of functions (injective, surjective, bijective). The focus is on the interpretation of a group as a set of mappings that operate on a space, without neglecting the general case of an abstractly given group. Secondly, the book is intended to introduce the modern mathematical developments of group theory. Here too, the approach is as elementary as possible, but, of course, it is unavoidable that the material in the later chapters will be more difficult to read than that in the earlier chapters. Nevertheless, some of the central mathematical concepts are so elementary that they can be seamlessly connected to the first chapters. Thirdly, one learns some things about geometry in this book. The structure of the Euclidean and hyperbolic plane as well as that of the 2-sphere is examined through their isometries. This results in a deeper understanding of the symmetry properties of known objects, such as, for example, the regular polyhedra. Finally, many examples are implemented with the freely available group theory program GAP. For many beginners, groups are something so abstract that any concretization can help them. GAP as well as geometric visualizations provide the requisite tangible elements. With the advancing computer technology for mathematically interested people, it certainly makes sense to learn how to handle an algebra program—this happens, essentially as a side-effect, when reading this text.

The book is aimed at Bachelor's, Master's and teaching students of mathematics and the natural sciences who are coming into contact with group theory for the first time. It is also intended for students and scientists who want to delve into the modern geometric aspects of group theory.

The first chapter deals with geometry, as far as it is necessary for the further chapters. It is about isometries, their composition, permutations and associated notation. In the second and third chapters, the basics of group theory are laid out as geometrically as possible. This part is especially suitable for self-study, and does not require an accompanying lecture.

In the fourth chapter, permutations are considered abstractly for the first time, without reference to geometry, with the introduction of the symmetric and alternating groups. Here is formalized what was done geometrically in Chaps. 2 and 3: groups operate on sets. In Sect. 4.6, the theorem of Švarc–Milnor is explained using an example. A precise formulation of this theorem can be found in Chap. 10. Here, as in many other places, a central methodological concept of the book becomes clear: The ideas behind the relevant facts are presented using examples before giving a precise mathematical formulation.

In Chap. 5, the presentation of a group by generators and relations is treated. Closely related to this are the decision questions formulated by Max Dehn at the beginning of the twentieth century.

In the following sixth chapter, products of groups are discussed. Presentations of direct, free and semidirect products of groups are developed. The chapter ends with an overview of the translation subgroups of discontinuous groups of the plane.

Chapter 7 deals with finite groups. The classical Sylow theorems are, following the theme of the book, proven using operations of groups. Finite groups of (very) small order are classified up to isomorphism. In Sect. 7.5, the regular decompositions of the 2-sphere are systematically analyzed.

Chapter 8 deals with abelian and solvable groups. Finitely generated abelian groups are classified and series of groups are studied.

Chapter 9 is dedicated to hyperbolic geometry and the groups that operate on the hyperbolic plane. The axiomatic approach to geometry is briefly discussed, then some intuitive geometric concepts of hyperbolic geometry are generated for the readers. Finally, decompositions of the hyperbolic plane are considered and presentations of their symmetry groups are given.

In Chap. 10, the concepts and insights gained from the previous chapter are applied and expanded on the Cayley graph of a group. Here, the idea of quasi-isometry and hyperbolic groups is explained and one of the essential theorems from the theory of hyperbolic groups is proven, namely the existence of a linear isoperimetric inequality for hyperbolic groups. By the end of the book, although the proofs of some of the tools are absent (their inclusion would have exceeded the scope of the project), the guiding ideas and methods will have been clearly demonstrated.

Exercises appear at the end of most sections. There, readers are encouraged to deepen their understanding by actively implementing the gained insights. Hints for solutions, sometimes even the complete solutions, are listed in the appendix. Of course, readers will spoil their chance for self-evaluation if they look at the back without first tackling the exercises independently! If you notice

errors, deficiencies, or suggestions for improvement, please send a message to
rosebrock@ph-karlsruhe.de.

In conclusion, I feel the need to express my thanks to several people. First and
foremost, I would like to mention Dr. Cynthia Hog-Angeloni. Without her help,
I would not have been able to realize the book at hand. She has accompanied
me from the beginning of my project. Dr. Hog-Angeloni has discussed the topic,
its conceptualization and organization with me for each chapter and subsequently
found many errors. I thank her particularly emphatically. Ms. Anna Schill read the
first three chapters and worked her way into an unfamiliar area. She pointed out
many places that could be made more readable with her help. I also thank Ulrike
Krell for creating Figs. 7.4 and 9.10, Hanno Rehn for Fig. 9.9, and Saskia Rosebrock
for assisting in the post-processing of the figures, Prof. Dr. Jens Harlander, Prof. Dr.
Wolfgang Metzler, Stephanie Ginaidi and Holger Blasum for content suggestions,
and Prof. Dr. Cornelia Rosebrock for linguistic corrections.

Karlsruhe, Germany                                                          Stephan Rosebrock

# Contents

# Chapter 1
# Introduction to Euclidean Geometry

The first chapter is not about groups yet. Here, familiar concepts from school, such as reflections and rotations, are revisited and formalized. In addition, such maps are composed so that we can calculate with them. Additionally, we prove a few theorems of Euclidean geometry that clarify the structure of the Euclidean plane and higher-dimensional spaces. We want to understand geometry here in a way that allows us to grasp it algebraically.

## 1.1 Isometries

Imagine we are holding an equilateral triangle made of wood. If we rotate it by $120°$ or by $240°$ then it will end up in exactly the same position in our hand. But we can also flip it, holding one of the three corners. So flipping is possible in three ways, depending on which corner we hold.

A bit more formally: How can we map an equilateral triangle onto itself? There is the identity $id$, which maps each point onto itself. We can also reflect the triangle across a line which passes through a vertex and the center of the opposite edge (this corresponds to the flipping above). There are three of these lines, labeled $a, b, c$ in Fig. 1.1.

For example, if we reflect over $a$, the vertex 1 is mapped to the vertex 2, and 3 is mapped onto itself. The map given by reflection over $a$ is called $s_a$. So it follows that $s_a(1) = 2$, $s_a(2) = 1$ and $s_a(3) = 3$.

There are more maps of this triangle onto itself. We can rotate the triangle about its center counterclockwise, either by $120°$ or by $240°$. Let's call these rotations $d_{120}$ and $d_{240}$. There are no further rotations of the triangle: A rotation by $360°$ is the same as a rotation by $0°$, because we are considering maps of the triangle onto itself, and as a map, a rotation by $0°$ is the same as a rotation by $360°$. In both maps

© The Author(s), under exclusive license to Springer-Verlag GmbH, DE, part of Springer Nature 2024
S. Rosebrock, *Visual Group Theory*, Springer Undergraduate Mathematics Series, https://doi.org/10.1007/978-3-662-69365-0_1

**Fig. 1.1**  An equilateral
triangle

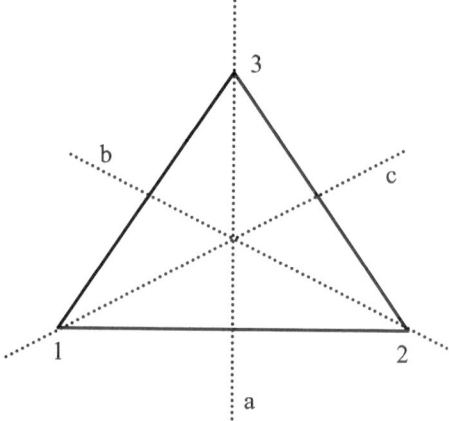

the vertex 1 is mapped to the vertex 1, 2 to 2 and 3 to 3. The rotation by $0°$ is the
same as the identity *id*.

So we get for the set of maps of the regular triangle onto itself:

$$D_3 = \{id, d_{120}, d_{240}, s_a, s_b, s_c\}.$$

There are no more maps of the triangle onto itself: In each such map, a vertex of
the set $\{1, 2, 3\}$ is mapped onto a vertex of the same set. The vertices are therefore
permuted. But there are only 6 permutations of 3 vertices and a given permutation
of the vertices determines the entire map.

The maps of $D_3$ not only map the triangle onto itself, but even the whole plane
onto itself, if we think of the triangle as being given in the plane. All maps in $D_3$
are *distance-preserving*: If two points $A$ and $B$ distance $r$ apart are mapped by an
element $f \in D_3$, then $f(A)$ and $f(B)$ are also distance $r$ apart. Every line segment
is mapped onto a line segment of the same length by a distance-preserving map.

Here we refer to the *plane* as the *Euclidean plane*, i.e., the set of all points of
$\mathbb{R} \times \mathbb{R} = \{(x, y) \mid x, y \in \mathbb{R}\}$, in which distances can be measured according to the
Pythagorean theorem: For points $P_1 = (x_1, y_1)$ and $P_2 = (x_2, y_2)$ their distance is
defined as

$$d_e(P_1, P_2) = \sqrt{(x_1 - x_2)^2 + (y_1 - y_2)^2}.$$

**Definition 1.1**  A (planar) *isometry* is a distance-preserving bijective map of the
plane onto itself.

Isometries not only preserve distances, but also angles, as one can easily see.
Isometries always map straight lines onto straight lines.

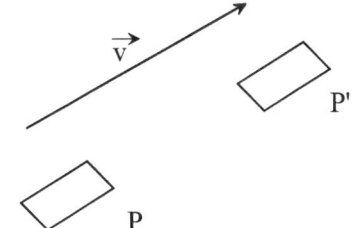

**Fig. 1.2**  Translation

**Fig. 1.3**  Frieze

Typical examples of isometries are the following:

- Given a fixed straight line $a$ in the plane, then the *reflection* $s_a$ over $a$ is an isometry.
- A *rotation* about a certain point by a certain angle counterclockwise is an isometry. In every rotation, the distances between two points are preserved.
- Similarly, a *translation*, i.e., a shift of the plane by a fixed amount in a fixed direction, is an isometry. Every point of the plane is moved by the same amount in the same direction. A translation can be described by a vector whose length is the amount of the translation and whose direction indicates the direction of translation.

    In Fig. 1.2, the vector $\vec{v}$ shifts the entire plane in the given direction by its length. Under this map $t_v$, for example, the parallelogram $P$ is mapped onto $P'$.
- A *glide reflection* is a reflection followed by a translation in the direction of the mirror axis. This mirror axis is called the *glide reflection axis* or *axis of the glide reflection*. If one imagines the *frieze* (i.e., an infinitely long, constantly repeating figure—a precise definition can be found on page 118) of Fig. 1.3 extended infinitely in both directions, then the mapping which consists of reflection over $a$ followed by translation by the vector $\vec{v}$ is a glide reflection that maps the frieze onto itself.

**Example 1.2** We consider a decomposition of the plane into squares as shown in Fig. 1.4 (which should be thought of as extending infinitely in all directions). Vertical and horizontal translations by integer multiples of the square width are *symmetries* of this decomposition, thus mapping the decomposition onto itself.

Further such symmetries are reflections through all drawn straight lines, all diagonals and all perpendiculars to the midpoints of the squares. Rotations of 90, 180 and 270° can be performed around all the intersection points of straight lines

**Fig. 1.4** Decomposition of
the plane into squares

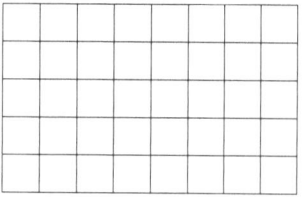

and around the centers of squares. Around the centers of each edge of a square, one can rotate by 180°. There are also glide reflections: The corresponding axes run through the centers of adjacent edges of a square.

Of course, the *identity*, which we denote by *id*, i.e., the map that maps every point of the plane onto itself, is an isometry. We can also regard *id* as a rotation by 0° (the *trivial rotation*) or as a translation with a translation vector of length 0 (the *trivial translation*). Therefore, we do not need to list the identity separately. Also, a *point reflection*, which is often treated in school, is an isometry of the plane. It will not be specifically mentioned here since a reflection through a point $P$ in the plane is the same as a rotation of 180° around $P$.

In the appendix, it is proven that every planar isometry is of one of the above-mentioned 4 types (translation, rotation, reflection, and glide reflection).

**Exercises**

1. Which isometries map a decomposition of the plane into congruent equilateral triangles onto itself?
2. Describe all isometries of a regular $n$-gon in the plane.
3. Draw a frieze that allows a reflection, but no rotations.

## 1.2   Figures and Permutations

**Definition 1.3** A *figure* is a subset of the plane.

For example, a *regular $n$-gon*, i.e., an $n$-gon with equal interior angles and equal edges, is a figure. Likewise, the frieze of Fig. 1.3 is a figure.

**Definition 1.4** An *isometry* of a figure is an isometry of the plane that maps the figure onto itself.

Above, we have already found the set of isometries of a regular triangle: $D_3 = \{id, d_{120}, d_{240}, s_a, s_b, s_c\}$ with the identity, 2 rotations, and 3 reflections.

Every figure has the identity *id* as an isometry. There are figures that do not allow any further isometries. The more isometries a figure allows, the more symmetrical it appears to the viewer. In this sense, the circle is very symmetrical. Every line through the center of the circle is an axis of reflection. Moreover, there are rotations by any angle around the center of the circle.

**Fig. 1.5** Rhombus

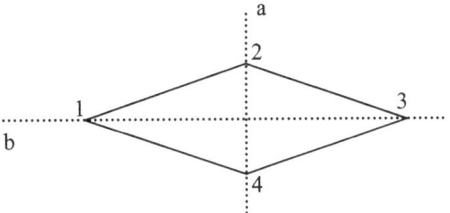

As another example, we consider the rhombus of Fig. 1.5. Apart from the identity, we can reflect over $a$ or $b$ and rotate the rhombus around its center by 180°, i.e.:

$$R = \{id, s_a, s_b, d_{180}\}.$$

There is a practical notation for many isometries in the plane. To introduce this, we observe that for every isometry of a regular $n$-gon (and many other figures as well) the entire map is determined solely by specifying the images of the vertices. For example, the mapping $s_a \in D_3$ maps point 1 to point 2 and point 2 to point 1 and leaves 3 fixed (see Fig. 1.1).

We use the *permutation notation*, in our example $s_a \simeq (1, 2)(3)$. In each pair of brackets, a point is mapped to the point that follows it. The brackets are read cyclically. So the last number in a pair of brackets is mapped to the first. The first bracket in $s_a \simeq (1, 2)(3)$ ensures that 1 is mapped to 2 and 2 to 1, and the second bracket means that 3 remains fixed. The reflection over the line $a$ is of course not the same as the permutation $(1, 2)(3)$, in which only numbers are swapped. The symbol $\simeq$ is meant to suggest that $(1, 2)(3)$ describes the reflection $s_a$. This is of course only the case if the vertices of the triangle are numbered as in Fig. 1.1.

For the identity, $id \simeq (1)(2)(3)$ applies. We have $d_{120} \simeq (1, 2, 3)$, so 1 is mapped to 2, 2 to 3 and 3 to 1. Instead of $s_b \simeq (1, 3)(2)$ we write $s_b \simeq (1, 3)$, leaving out the vertices that remain fixed. With this shorthand notation, we are allowed to write $id \simeq ()$.

Since the brackets are to be read cyclically, we have $(1, 2, 3) = (3, 1, 2) = (2, 3, 1)$. We can therefore describe the isometries of the regular triangle as follows:

$$D_3' = \{(), (1, 2, 3), (1, 3, 2), (1, 2), (1, 3), (2, 3)\}.$$

There is an analogous description of the isometries of the rhombus::

$$R' = \{(), (1, 3), (2, 4), (2, 4)(1, 3)\}.$$

The rotation by 180° around the center of the rhombus can be described by $(2, 4)(1, 3)$, which should not be confused with $(2, 4, 1, 3)$. The last permutation does not come from an isometry of the rhombus. A *permutation* is a bijective map of a set onto itself, and the permutation notation can be used to describe permutations of finite sets.

We now want to combine isometries and study properties of this combination. The *composition* of two distance-preserving bijections is again a distance-preserving bijection. So if we compose two elements of $D_3$ (i.e., we execute the corresponding maps successively), a third element of $D_3$ must result. It is said that *the set $D_3$ is closed with respect to composition.* Similarly, the addition of integers is closed, because the sum of two integers is again an integer.

The composition of isometries is often described using the permutation notation. As an example, we first execute $s_a$ and then $s_b$, i.e., the isometry $s_b \circ s_a$. We follow the images of the vertices under this isometry:

$s_a(1) = 2$ and $s_b(2) = 2$. So $s_b \circ s_a(1) = 2$.
$s_a(2) = 1$ and $s_b(1) = 3$. So $s_b \circ s_a(2) = 3$.
$s_a(3) = 3$ and $s_b(3) = 1$. So $s_b \circ s_a(3) = 1$.
Overall, $s_b \circ s_a \simeq (1, 2, 3) \simeq d_{120}$.

The isometries $s_a$ and $s_b$ can also be written in their permutation notation and directly follow the images of the vertices: $s_b \circ s_a \simeq (1, 3) \circ (1, 2) = (1, 2, 3)$.

Here, some caution is required: We adhere to the convention of evaluating products from the right, so $y \circ x$ means first $x$ and then $y$. Both reading directions can be found in the literature, and both have advantages and disadvantages. The advantage of our convention is that we can compose isometries in the same way as we would with functions.

A common question is whether the mirror axes themselves are mirrored. This is not the case. So in the isometry $s_b \circ s_a$, the mirror line $b$ remains in its original place.

Some more examples:

$$d_{120} \circ d_{120} = d_{120}^2 \simeq (1, 2, 3) \circ (1, 2, 3) = (2, 1, 3) \simeq d_{240};$$

$$d_{120}^3 = id;$$

$$s_a \circ s_b \simeq (1, 2) \circ (1, 3) = (1, 3, 2) \simeq d_{240} \neq s_b \circ s_a.$$

The composition of isometries is therefore generally not commutative! This is different from the addition or multiplication of integers, where the commutative law always applies.

The set of isometries of a given figure is closed with respect to their composition: An isometry that leaves a given figure fixed, composed with another isometry that leaves the same figure fixed, leaves the figure fixed overall.

The identity *id* behaves with respect to composition like 0 with respect to addition: If you compose any isometry $g$ with the identity, it gives us $g$ again. The identity *id* is the *identity element* with respect to composition.

Let's compose a reflection with the rotation of the rhombus:

$$s_a \circ d_{180} \simeq (1, 3) \circ (2, 4)(1, 3) = (2, 4)(1)(3) = (2, 4) \simeq s_b.$$

For every isometry, the map that undoes this isometry is bijective and distance-preserving and therefore also an isometry. Thus, for every isometry, there is another one, called the *inverse isometry*, such that the composition of the two isometries results in the identity.

To undo a reflection one reflects over the same line again. So in $D_3$ we have $s_a \circ s_a = s_a^2 = id$. We have $d_{120} \circ d_{240} = id$. The inverse of the isometry $f$ is denoted $f^{-1}$. So $s_a^{-1} = s_a$ and $d_{120}^{-1} = d_{240}$.

Regarding addition, there are no inverses in the natural numbers. Given a natural number, one cannot find another one such that their sum becomes 0. In the integers, however, there are inverses: The inverse of 7 is $-7$ and the inverse of $-26$ is 26.

**Exercises**

1. Which isometries do a rectangle and a parallelogram allow?
2. What is the result of the composition of two reflections with perpendicular mirror axes $a$, $b$? Is it the case that $s_a \circ s_b = s_b \circ s_a$?
3. Draw an infinite figure that allows a translation, but no reflection or glide reflection as an isometry.

## 1.3   The Structure of Isometries

Let $F_n$ be a regular $n$-gon in the plane (e.g. the regular triangle of Fig. 1.1 or the square of Fig. 1.6). Which isometries does $F_n$ allow?

Every figure has the identity as an isometry. The $n$-gon allows rotations by $k \cdot 2\pi/n$ for $k$ from 1 to $n-1$. If we consider the identity as a rotation by $0°$, we have rotations by

$$0, \quad \frac{2\pi}{n}, \quad 2\frac{2\pi}{n}, \quad 3\frac{2\pi}{n}, \dots, \quad (n-1)\frac{2\pi}{n}.$$

Of course, we can continue to rotate, but we do not get any new isometries. A rotation by $n * 2\pi/n = 2\pi$ corresponds to the identity, and a rotation by $(n+3) * 2\pi/n$ corresponds to the same map as a rotation by $3 *$

**Fig. 1.6**   A square

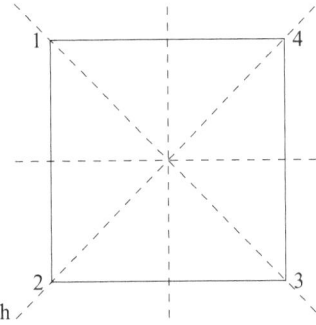

$2\pi/n$. The 4 rotations of the square of Fig. 1.6 are in permutation notation $\{(), (1, 2, 3, 4), (1, 3)(2, 4), (1, 4, 3, 2)\}$.

All rotations preserve the *orientation*, i.e., they preserve the cyclic order of the vertices, read counterclockwise. Except for the cyclic reading, the 4 vertices of the square after a rotation are still 1 2 3 4. The square is not "flipped" during a rotation. Rotations as well as translations are therefore *orientation-preserving isometries*. Orientation-preserving isometries preserve a turning direction in the plane.

A regular $n$-gon has more isometries, namely $n$ reflections. If $n$ is odd, as in the case of an equilateral triangle, the axes of reflection each run through a vertex and the midpoint of the opposite edge. If $n$ is even, as in the case of a square, then $n/2$ axes of reflection run through opposite vertices and $n/2$ axes of reflection run through opposite midpoints of the edges. Reflections are always *orientation-reversing*, they change the cyclic order of the vertices when read counterclockwise. The figure, and thus the entire plane, is "flipped". In a reflection, a choice of direction of rotation changes to the opposite direction of rotation. In Fig. 1.7, the word "Auto" is reflected over the line $a$, and thus the indicated direction of rotation changes.

In total, the set of isometries of a regular $n$-gon consists of $2n$ elements. If $d$ denotes the rotation by $2\pi/n$ and $s_1, \ldots, s_n$ the reflections, then $F_n$ has the isometries:

$$D_n = \{id, d, d^2, \ldots, d^{n-1}, s_1, \ldots, s_n\}.$$

(Be careful: Some authors call this set $D_{2n}$.)

There cannot be more isometries of a regular $n$-gon. We obtain each isometry of a regular $n$-gon in the following way: We can map the vertex 1 to $n$ different vertices, resulting in $n$ possibilities. For the vertex 2, we only have the option of mapping it to the right or left of the image vertex of 1, resulting in a total of $2n$ possibilities. But then the mapping is fixed, so there can be no more than $2n$ isometries of a regular $n$-gon.

The composition of two orientation-preserving isometries is orientation-preserving. If two orientation-reversing isometries are executed one after the other, the orientation has been reversed twice. So overall, an orientation-preserving isometry is obtained.

**Fig. 1.7** Reflection

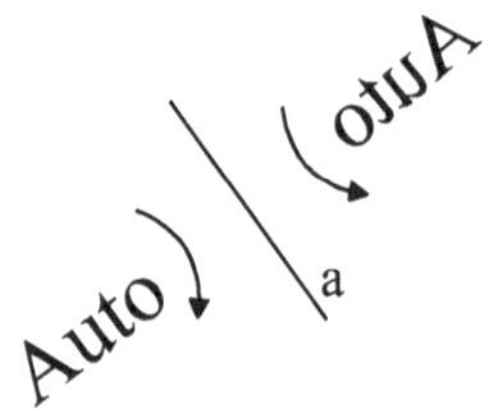

**Definition 1.5** If $f$ is an isometry in the plane $\mathbb{R}^2$, then any point $P \in \mathbb{R}^2$ for which $f(P) = P$ holds is called a *fixed point* of $f$. The isometry $f$ is called *fixed-point-free* if for all points $P$ of the plane we have $f(P) \neq P$.

If an isometry $f$ maps a figure $F$ onto itself (i.e., $f(F) = F$), then $F$ is called *invariant* under $f$. Be careful, in this case $f$ does not have to have any fixed points on the figure $F$, as for example in a 120° rotation of the equilateral triangle about its center.

**Theorem 1.6** *An isometry of the plane with a fixed point is a rotation if it preserves the orientation. It is a reflection if it does not preserve the orientation. An isometry of the plane without a fixed point is a translation if it preserves the orientation, and otherwise a glide reflection.*

The proof of this theorem can be found in the appendix and can be skipped on first reading. Rotations, reflections, translations, and glide reflections are thus all occurring isometries in the plane.

Three points in the plane are called *collinear* if they lie on a straight line.

**Theorem 1.7** *Any isometry of the plane that has 3 non-collinear fixed points is the identity.*

**Proof** Let $P, Q, R$ be non-collinear fixed points of the isometry $f$. Since $f$ fixes the points $P$ and $Q$ and preserves lengths, it must also fix the line $l$ through $P$ and $Q$. Thus, $f$ is either the identity or a reflection in $l$. However, the reflection does not have the fixed point $R$, because $R$ does not lie on $l$. □

**Corollary 1.8** *A plane isometry is uniquely determined by the image of three non-collinear points.*

**Proof** Let $g, h$ be two isometries such that $g(P) = P'$, $g(Q) = Q', g(R) = R'$ and $h(P) = P'$, $h(Q) = Q'$, $h(R) = R'$ for non-collinear points $P, Q$ and $R$. The isometries $g$ and $h$ can differ at most by an isometry $f$ that fixes $P, Q, R$, so $g = h \circ f$ and $f(P) = P, f(Q) = Q, f(R) = R$. According to Theorem 1.7, $f$ must be the identity, so that $g = h$ follows. □

**Theorem 1.9** *The product of two reflections along parallel lines is a translation perpendicular to the mirror axes by twice the distance between them. The product of two reflections along mirror axes that intersect at a point $P$ is a rotation about this point by twice the angle between the two mirror axes.*

**Proof** Let $a$ and $b$ be two parallel mirror axes. The isometry $t = s_a \circ s_b$ leaves all lines that are perpendicular to $a$ and $b$ invariant, and therefore $t$ is not a rotation. $t$ is orientation-preserving as the product of two orientation-reversing isometries, and thus $t$, because it is not a rotation, is a translation according to Theorem 1.6. Every point on $a$ is shifted by the isometry $s_b$, and thus by $t$, by twice the distance between $a$ and $b$. If this applies to points on $a$, it naturally also applies to all other points.

Now let $a$ and $b$ be two mirror axes that intersect in a point $P$. Every circle with center $P$ remains invariant under $s_a$ as well as under $s_b$ and thus also under $d = s_a \circ$

$s_b$. The isometry $d$ is orientation-preserving as the composition of two orientation-reversing isometries and thus a rotation about $P$ according to Theorem 1.6. If one follows a point on $a$ under $d$, one sees that the rotation angle corresponds exactly to twice the angle between the mirror axes.                                                  □

Given a set and an operation we can generate the corresponding *multiplication table*.

**Example 1.10** This is the multiplication table for the set $D_3$ with respect to composition:

| ∘ | $id$ | $d_{120}$ | $d_{240}$ | $s_a$ | $s_b$ | $s_c$ |
|---|---|---|---|---|---|---|
| $id$ | $id$ | $d_{120}$ | $d_{240}$ | $s_a$ | $s_b$ | $s_c$ |
| $d_{120}$ | $d_{120}$ | $d_{240}$ | $id$ | $s_b$ | $s_c$ | $s_a$ |
| $d_{240}$ | $d_{240}$ | $id$ | $d_{120}$ | $s_c$ | $s_a$ | $s_b$ |
| $s_a$ | $s_a$ | $s_c$ | $s_b$ | $id$ | $d_{240}$ | $d_{120}$ |
| $s_b$ | $s_b$ | $s_a$ | $s_c$ | $d_{120}$ | $id$ | $d_{240}$ |
| $s_c$ | $s_c$ | $s_b$ | $s_a$ | $d_{240}$ | $d_{120}$ | $id$ |

For example we see here that $s_b \circ d_{120} = s_a$ because the entry in the line $s_b$ and the column $d_{120}$ is $s_a$.

**Exercises**

1. Does $(1, 3)(2, 4)$ (or $(1, 2, 3)$) describe an isometry of the square of Fig. 1.6? If yes, which one? If no, why not?
2. Generate a glide reflection by executing three reflections. (Hint: Generate a translation through two reflections according to Theorem 1.9.)
3. Create a multiplication table for the symmetries of the rhombus in Fig. 1.5 with respect to composition.

## 1.4   Higher-Dimensional Spaces

So far, we have only considered isometries in the plane. Similarly, one can consider isometries in higher-dimensional Euclidean spaces. An *isometry* is a distance-preserving bijective mapping of $\mathbb{R}^n$ onto itself. For $n = 2$, we get the original definition of plane isometry. In $\mathbb{R}^3$, for example, we can reflect through a plane or rotate around a line.

Let's consider the isometries of the cube of Fig. 1.8. For example, we can rotate around the line $a$ by 90, 180 and 270° clockwise. The rotation by 90° has the following form in permutation notation: $(1, 5, 6, 2)(4, 8, 7, 3)$. There are three rotation axes of this type. There are also rotations around lines through the centers of opposite cube edges by 180°. For example, if we rotate about the line through

**Fig. 1.8** Cube

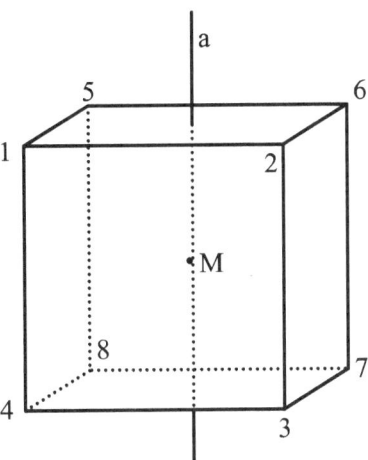

the centers of edges 2,6 and 4,8 by 180°, we get the rotation $(2, 6)(4, 8)(5, 3)(1, 7)$. There are six rotation axes of this type.

But there are also rotations by 120 and 240° each about one of the 4 diagonals of the cube. One example is the rotation $(4, 2, 5)(3, 6, 8)$. The associated rotation axis goes through the vertices 1 and 7.

We count all rotations: The identity, 9 rotations about lines through side centers, 8 rotations about diagonals and 6 rotations about lines through opposite edge centers together result in 24 orientation-preserving isometries.

We can reflect through the plane passing through the vertices 2,3,5,8 and get the reflection $(1, 6)(4, 7)$. There are six such mirror planes on which opposite edges of the cube lie. A plane parallel to a cube side that goes through the center of the cube is also a mirror plane. $(1, 4)(2, 3)(5, 8)(6, 7)$ describes a corresponding reflection. There are three of this type of reflection.

In addition, the cube allows a *point reflection* $s_M$. In this case, each point $P \in \mathbb{R}^3$ is reflected through the center $M$ of the cube in the following way: $s_M(P)$ lies on the line through $P$ and $M$ so that $M$ is exactly in the middle between $P$ and $s_M(P)$. The point reflection through the cube center can thus be described by $(1, 7)(5, 3)(6, 4)(2, 8)$.

Further orientation-reversing isometries of the cube are not so easy to imagine. But if you compose a rotation with $s_M$, you get an orientation-reversing isometry. So there are also 24 of these.

In general, a reflection in $\mathbb{R}^n$ can be performed through an $n - 1$-dimensional *hyperplane*, which is a copy of $\mathbb{R}^{n-1}$ in $\mathbb{R}^n$. Another example is a rotation around an $n - 2$-dimensional hyperplane (a line in $\mathbb{R}^3$ or a point in the plane). But you can also reflect through a line in $\mathbb{R}^3$. This isometry corresponds to a rotation around the line by 180°. If you reflect through $a$ in Fig. 1.8, you get the isometry described by $(1, 6)(2, 5)(4, 7)(3, 8)$.

There are analogous theorems to Theorem 1.7 and Corollary 1.8:

**Theorem 1.11** *An isometry in $\mathbb{R}^n$ that fixes $n + 1$ linearly independent points is the identity.*

**Proof** For $n = 2$, the assertion corresponds to Theorem 1.7. Let's assume inductively that every isometry in $\mathbb{R}^{n-1}$ that fixes $n$ linearly independent points is the identity. Let $n + 1$ linearly independent points $P_1, \ldots, P_{n+1}$ in $\mathbb{R}^n$ and an isometry $f$ be given with $f(P_i) = P_i$. By the induction assumption, $f$ maps the $n - 1$-dimensional hyperplane $W$, in which the points $P_1, \ldots, P_n$ lie, identically. So $f$ is either a reflection through $W$ or the identity. But the reflection through $W$ would not have $P_{n+1}$ as a fixed point.                                     □

The proof of the following corollary corresponds to that of Corollary 1.8.

**Corollary 1.12** *An isometry in $\mathbb{R}^n$ is uniquely determined by the image of $n + 1$ linearly independent points.*

**Exercises**

1. Which isometries do a sphere and a tetrahedron in $\mathbb{R}^3$ allow?
2. Calculate the result of the composition of the rotation described by $(4, 2, 5)(3, 6, 8)$ and the point reflection through the center of the cube.
3. Prove:

   (a) Every isometry of $\mathbb{R}^n$ can be represented by a composition of translations, rotations and reflections.
   (b) Every isometry of $\mathbb{R}^n$ can be represented by a composition of reflections.

# Chapter 2
# Introduction to Groups

In this chapter, we define the concept of a group. Our examples of groups are mostly, but not always, the set of all symmetries of a given figure. We learn the first important properties of groups and how we can "generate" a group by using only a few elements. We thus answer questions like: *How do I get all other isometries of a figure from a few?* Then we study the groups generated by one element. We look a little more closely at all the symmetries of the tetrahedron towards the end of this chapter.

Although GAUSS already implicitly used groups (he introduced an operation on quadratic forms, which thus form a group) and important theorems of group theory were proven even earlier (e.g., by LAGRANGE and EULER), groups were only explicitly used by the brilliant French mathematician ÉVARISTE GALOIS [1811–1832]. The problem of the algebraic solution of equations was completely solved by him using groups. In 1815, AUGUSTIN-LOUIS CAUCHY [1789–1857] was the first to systematically study groups. In his case, they were groups of permutations. ARTHUR CAYLEY [1821–1895] was the first to introduce abstract groups in 1854.

## 2.1 The Definition of a Group and the Dihedral Groups

In the last chapter, we considered the set $D_3 = \{id, d_{120}, d_{240}, s_a, s_b, s_c\}$ of isometries of a regular triangle in the plane. Here, $s_a$, $s_b$ and $s_c$ are reflections over axes $a$, $b$ and $c$. The isometries $d_{120}$ and $d_{240}$ are rotations by 120 and 240° as shown in Fig. 1.1. $id$ is the identity. We also found that the isometries of $D_3$ are closed with respect to composition $\circ$. The set $D_3$ with the operation of composition is a typical example of a group.

S. Rosebrock, *Visual Group Theory*, Springer Undergraduate Mathematics Series, https://doi.org/10.1007/978-3-662-69365-0_2

**Definition 2.1** Let $G$ be a set and $\cdot$ an operation, with respect to which $G$ is closed. The pair $(G, \cdot)$ is called a *group* if it fulfills the following properties:

1. (Associativity) For all $u, v, w \in G$:

$$(u \cdot v) \cdot w = u \cdot (v \cdot w). \tag{2.1}$$

2. (Existence of an identity element) There is an $e \in G$ such that

$$e \cdot g = g \cdot e = g, \quad \forall g \in G. \tag{2.2}$$

$e$ is called the *identity element* of the group $G$.

3. (Existence of inverse elements) For each $g \in G$ there is a $g' \in G$ such that

$$g \cdot g' = g' \cdot g = e, \tag{2.3}$$

where $e$ is the identity element. $g'$ is called the *inverse* of $g$.

We will prove in Theorem 2.17 on page 24 that there is exactly one identity element in every group. However, it is more important to us at the moment to gain an understanding of groups.

We want to follow our considerations on a computer. For this we use the freely available software package GAP (see [GAP22]). We can define a group by specifying its elements in the permutation notation:

```
gap> D3:=Group((),(1,2,3),(1,3,2),(1,2),(1,3),(2,3));
Group([ (), (1,2,3), (1,3,2), (1,2), (1,3), (2,3) ])
```

Inputs are made in GAP after the prompt gap>. The second line is the output.

Let's take another look at our pair $(D_3, \circ)$. We will soon prove that $(D_3, \circ)$ is a group. For example, the identity element is the identity *id*.

Indeed, $g \circ id = g$ holds for all $g \in D_3$ (here for example $g = (1, 2)$).

```
gap> ()*(1,2);
(1,2)
```

In GAP, the composition symbol $\circ$ is given by the symbol $*$.

The inverse of $d_{120}$ is the element $d_{240}$ (i.e. $d'_{120} = d_{240}$). If you compose (execute one after the other) $d_{120}$ with $d_{240}$, you get the identity:

```
gap> (1,3,2)*(1,2,3);
()
```

In GAP elements are concatenated from left to right, while we concatenate from right to left.

A reflection is undone by reflecting again over the same axis, i.e. $s_a \circ s_a = s_a^2 = id$ or, in other words, $s_a = s'_a$. A nontrivial map that is inverse to itself is called an involution. Or in general: An element of a group that is not the identity, but whose square is the identity, is called an *involution*.

A reflection is inverse to itself:

Let's check the associativity with an example:

```
gap> (1,2)*(1,2);
()
```

$$(d_{240} \circ s_c) \circ s_a = d_{240} \circ (s_c \circ s_a)$$

$$\Leftrightarrow \quad (1,3) \circ s_a = d_{240} \circ (1,3,2)$$

$$\Leftrightarrow \quad (1,2,3) = (1,2,3).$$

The same in GAP:

```
gap> ((1,2)*(2,3))*(1,3,2);
(1,2,3)
gap> (1,2)*((2,3)*(1,3,2));
(1,2,3)
```

More generally, we have the following fact:

**Example 2.2** For $n \geq 2$, $(D_n, \circ)$ forms a group, the so-called *dihedral group*.

**Proof** If you perform two distance-preserving maps one after the other, each of which leaves a figure fixed, then you will again get a distance-preserving map which leaves the same figure fixed. This proves the closure. We check the associativity (2.1) of Definition 2.1: Let $x$ be any point of the plane. Then for all $u, v, w \in D_n$: $(u \circ v) \circ w(x) = u(v(w(x))) = u \circ (v \circ w)(x)$. Thus the concatenation of maps is always associative.

The identity is the identity element. What remains to check is the existence of the inverses (2.3): For a distance-preserving map $g$ which leaves a figure fixed, the map $g'$, which undoes $g$, is also distance-preserving and leaves the figure fixed. $g'$ is thus also an isometry of the same figure and thus an element of $D_n$.                □

Sometimes, as in the above proof, we write only $D_n$ instead of $(D_n, \circ)$, or just $G$ instead of $(G, \cdot)$, if there can be no misunderstandings regarding the operation.

Replacing a regular $n$-gon by any figure, we have shown even more with the last proof:

**Theorem 2.3** *The isometries of a figure in the plane form a group with respect to composition, the* symmetry group *of the figure.*

As a further example, we consider the group $D_4$ of the square (see Fig. 1.6). It consists of four reflections over the four marked axes and four rotations around the center of the square by the angles $0, 90, 180$ and $270°$, corresponding to the elements $id, d, d \circ d, d \circ d \circ d$.

Sometimes we write $d \circ d$ as $d^2$, with higher powers accordingly. Thus, the rotations of $D_4$ are written as $id, d, d^2, d^3$.

Let $G_{(4,4)}$ (the name will be explained later) be the symmetry group of the decomposition of the plane into squares of Example 1.2 on page 3. It consists of infinitely many elements. In each square of the decomposition, the eight symmetries of the square can be executed. Not only is the respective square mapped onto itself, but the whole decomposition.

As we justified in the proof of Example 2.2, the composition of isometries is associative. However, one can certainly imagine operations that are not associative:

**Example 2.4** We define the operation $a \diamond b = a + 2b$ on the integers. So, for example, $3 \diamond 4 = 11$ and $2 \diamond -3 = -4$.

This operation is not associative, because for example

$$(1 \diamond 2) \diamond 3 = 5 \diamond 3 = 11 \quad \text{but} \quad 1 \diamond (2 \diamond 3) = 1 \diamond 8 = 17.$$

Defining a group by writing down all its elements can be very tedious, and indeed it is impossible for groups with infinitely many elements. Instead, one would like to write only a set of elements that can generate all elements of the group by taking their products and inverses. For example, in the group of the square, there is no need to specifically write $d^2$ and $d^3$, because they can be generated from $d$.

**Definition 2.5** A group $G$ is *generated* by the elements $E = \{g_1, \ldots, g_n\}$ if every element of $G$ can be represented by combining the elements of $E$ and their inverses. The set $E$ is then called a *generating system* of the group $G$. Notation: $G = \langle g_1, \ldots, g_n \rangle$.

We prove with the help of GAP that the symmetry group of the regular triangle in the plane is generated by $s_a$ and $d_{120}$. In GAP, to define a group it is sufficient to specify the generators.

```
gap> D3:=Group((1,2),(1,2,3));
Group([ (1,2), (1,2,3) ])
gap> Elements(D3);
[ (), (2,3), (1,2), (1,2,3), (1,3,2), (1,3) ]
```

Indeed, we obtain all elements of $D_3$ in this way. We write $D_3 = \langle s_a, d_{120} \rangle$ to indicate that the group $D_3$ is generated by $s_a$ and $d_{120}$. It is not difficult to prove that the group $D_n$ is generated by a rotation of $360/n$ degrees and a reflection (see Exercise 5.).

In Sect. 1.4 we have already studied isometries in 3-dimensional space. Of course, every 3-dimensional object in $\mathbb{R}^3$ also has a symmetry group, just like in the planar case.

The symmetry group of the cube is generated by all rotations and the point reflection through the center of the cube. However, significantly fewer generators are sufficient, as we will justify in Sect. 7.5.

In the following GAP-Code we generate the group of the cube by three reflections through three different planes. The double semicolon prevents output in GAP.

```
gap> a:=(1,2)(5,6)(4,3)(8,7);; b:=(1,3)(5,7);;
     c:=(5,4)(6,3);;
gap> W:=Group(a,b,c);
Group([ (1,2)(3,4)(5,6)(7,8), (1,3)(5,7), (3,6)(4,5) ])
```

**Example 2.6** The symmetry group $G_{(4,4)}$ of the decomposition of the plane into squares consists of infinitely many elements. However, a finite number of elements is sufficient to generate the group.

In Theorem 4.31 on page 74 and Exercise 2. of Sect. 4.6 on page 88 it is shown that the three reflections $s_a$, $s_b$ and $s_c$ over the axes $a$, $b$ and $c$ of Fig. 2.1 generate the group $G_{(4,4)}$. The reflections $s_a$ and $s_b$ alone generate the group of that square in

**Fig. 2.1**  Generators of the group $G_{(4,4)}$

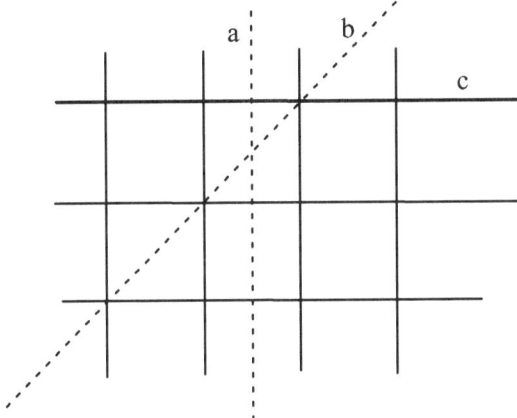

which the corresponding axes intersect. Together with $s_c$, all possible isometries are obtained. This can be seen (in the literal sense) if you place 3 mirrors on the axes $a, b, c$ of the (enlarged) Fig. 2.1 and look in from above.

**Exercises**

1. Write the elements of the group $D_4$ in the permutation notation. For each isometry of the square, indicate the inverse. (Hint: Use GAP to find a set of generators of the group $D_4$. You can also determine the inverses with GAP.)
2. Prove that $(\mathbb{Q}, +)$ forms a group.
3. Let $F$ be the set of lines in the plane described by the following functions:

$$F = \{f \mid f(x) = ax + b, \ a, b \in \mathbb{R}\}.$$

   Let $\cdot$ be the operation on $F$ defined by $f \cdot g = f(x) + g(x)$. Show that $(F, \cdot)$ is a group.
4. Define a "wild" operation on the integers (like the operation of Example 2.4), which is closed. Check whether the operation is associative and/or commutative. Is there exactly one identity element? Does your operation have an inverse for each element? Do the integers together with your operation form a group?
5. Prove that the group $D_n$ is generated by the rotation by $d = 360/n$ degrees and any reflection $s$. (Hint: The rotation $d$ generates all rotations. Compose $s$ with all rotations to get all reflections.)

## 2.2  The Order of a Group and Abelian Groups

**Definition 2.7**  The *order* of a group is the number of its elements.

On page 8 we justified that there are $2n$ maps of the regular $n$-gon onto itself. The group $D_n$ therefore has order $2n$ (one also writes $|D_n| = 2n$).

In our GAP input window we still have the group $W$ of the cube. Order gives us the group order:

```
gap> Order(W);
48
```

Since the elements $a, b, c$ from the above GAP code correspond to reflections of the cube of Fig. 1.8 on page 11 and the symmetry group of the cube has 48 elements (see Sect. 1.4 on page 10), GAP has shown that $a, b, c$ are generators of the cube group.

**Example 2.8** The symmetry group of a rhombus, which is not a square, is a group of order 4.

In the last chapter, we have exhibited the isometries of the rhombus. The symmetry group of the rhombus can then be written as

$$R = \{id, s_a, s_b, d_{180}\},$$

where $a$ and $b$ are perpendicular mirror axes through opposite vertices of the rhombus and $d_{180}$ is a $180°$ rotation around the center of the rhombus (see Fig. 1.5).

**Theorem 2.9** *Let $\mathcal{R}$ be the line, the plane, or the 3-dimensional space. Then the isometries of $\mathcal{R}$ form a group with respect to composition.*

**Proof** If you compose two distance-preserving maps, you get another distance-preserving map. This shows the closure of composition.

The associativity can be seen as in Example 2.2. The identity is the identity element. We still lack (2.3) of Definition 2.1: Given a distance-preserving map $g$, the map that undoes $g$ is also distance-preserving. Therefore, it is an isometry on $\mathcal{R}$.                                                                                        □

The groups mentioned in Theorem 2.9 all have infinite order. Similarly, the symmetry group of the frieze in Fig. 1.3 has infinite order, as it contains infinitely many translations. We introduce the notation $\mathcal{E}$ for the isometry group of the plane.

The elements of a group do not have to be maps, and the operation does not have to be composition. Group elements can be (among other objects) numbers, and the operation can be, for example, the normal addition or multiplication.

The natural numbers do not form a group with addition. There is not even an identity element. If we add 0 to the natural numbers, we have an identity element because $n + 0 = 0 + n = n$. But the inverses are missing. For this, we need the negative integers:

**Example 2.10** The integers with the usual addition $(\mathbb{Z}, +)$ form a group.

**Proof** The sum of two integers is again an integer. So the addition on the integers is closed.

The associativity can be easily seen geometrically. We need to show:

$$(n + m) + k = n + (m + k) \text{ for all integers } n, m, k.$$

If you join lines of lengths $n, m, k$ together end to end, in that order, you get a line of length $n + m + k$, regardless of whether you first combine the lines of lengths $n$ and $m$ into a unit and then add the line of length $k$, or whether you first combine the lines of lengths $m$ and $k$ and then add the line of length $n$.

The identity element is 0 and the inverse of an integer $n$ is $-n$.                    □

The inverse of 3 in $\mathbb{Z}$ is not written as $3'$ but, as usual, $-3$. So the minus sign is actually not an operation, as one learns in elementary school, but an inversion sign and more correctly one should write $5 + (-3)$ instead of $5 - 3$. If the group operation is multiplication or composition, the inverse of $g$ is written as $g^{-1}$.

We have $|\mathbb{Z}| = \infty$, because there are infinitely many integers.

We have seen in the first chapter that in the group $D_3$ for the reflections $s_a$ and $s_b$ we have

$$s_b \circ s_a \simeq (1, 3) \circ (1, 2) = (1, 2, 3) \simeq d_{120} \neq s_a \circ s_b.$$

So, the commutative law does not apply in the group $D_3$. In GAP:

```
gap> (1,3)*(1,2);
(1,3,2)
gap> (1,2)*(1,3);
(1,2,3)
```

However, for any two integers $n, m$, it always holds that $n + m = m + n$. This motivates the following definition:

**Definition 2.11** A group $(G, \cdot)$ is called *abelian* or *commutative* if for any two elements $g, h \in G$ it holds that $g \cdot h = h \cdot g$.

The groups $(D_n, \circ)$ are not abelian for $n \geq 3$: According to Theorem 1.9 on page 9, the product of two reflections is a rotation by the angle $2\alpha$ if the mirror axes intersect and form the angle $\alpha$ at the intersection point. The rotation is to be performed clockwise or counterclockwise, depending on which reflection is performed first.

The regular $n$-gon has two mirror axes $a, b$, which intersect at an angle of $180/n$ degrees. Their product is therefore a rotation by $360/n$ degrees. For $n = 2$, this

**Fig. 2.2** $s_a s_b$ and $s_b s_a$

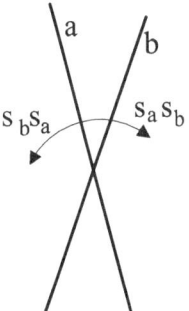

would be a rotation by $180°$. In that case, the same rotation would result, regardless of whether it was rotated clockwise or counterclockwise. However, for $n \geq 3$, it follows that $s_a s_b \neq s_b s_a$, since the clockwise rotation is different from the counterclockwise rotation (see Fig. 2.2).

If the group operation is $\circ$ or $\cdot$, we often omit the operator symbol, as here, so we write $gh$ instead of $g \cdot h$.

$(\mathbb{Z}, +)$ is abelian, because $n + m = m + n$ is true for all integers $n, m$.

Do the rational numbers form a group with the ordinary multiplication? The product of two rational numbers is rational (closure), the number 1 is the identity element. The associative law holds. The inverse of $a/b$ is $b/a$ for integers $a, b$. But wait, 0 has no inverse. Therefore:

**Example 2.12** $(\mathbb{Q} - \{0\}, \cdot)$ is an abelian group.

**Proof** It is easy to see that $(\mathbb{Q} - \{0\}, \cdot)$ is abelian: $a/b \cdot c/d = ac/bd = ca/db = c/d \cdot a/b$. Since we have removed zero from the rational numbers, we need to reconsider the closure: The product of two numbers from $\mathbb{Q} - \{0\}$ will never be zero and remains rational, i.e. is again in $\mathbb{Q} - \{0\}$.                                        □

Let $D_n^+$ be the set of rotations of $D_n$, including the element $id$ (which is a rotation by $0°$). As we have already seen, we have:

$$D_n^+ = \{id, d, d^2, d^3, \ldots, d^{n-1}\}.$$

These rotations form a group with respect to composition because if you compose two rotations, you get another rotation (closure), the identity as the identity element is present, and for each rotation $d^i$ there is another one such that their combination gives the identity. This is the rotation $d^{n-i}$, because $d^i \circ d^{n-i} = d^{i+n-i} = d^n = id$. The group $D_n^+$ is commutative, because $d^n d^m = d^{n+m} = d^{m+n} = d^m d^n$.

**Exercises**

1. Is the symmetry group of the rhombus abelian?
2. Prove that the set of pairs of integers

$$\mathbb{Z} \times \mathbb{Z} = \{(n, m) \mid n, m \in \mathbb{Z}\}$$

forms an abelian group with component-wise addition

$$(n, m) + (p, q) = (n + p, m + q).$$

3. Write the elements of the group $D_n^+$ in the permutation notation. Prove that the product of two arbitrary reflections of the group $D_n$ is an element of $D_n^+$. (Hint: use Theorem 1.9.)
4. Prove that the set of even integers $\{2n \mid n \in \mathbb{Z}\}$ forms a group with normal addition. Do the odd integers form a group with respect to addition?
5. Are dihedral groups abelian?

6. Prove that $\mathbb{Z} = \langle 1 \rangle$. It also holds that $\mathbb{Z} = \langle 2, 3 \rangle$, but $\mathbb{Z} \neq \langle 6, 15 \rangle$. Which sets of integers generate $\mathbb{Z}$?

## 2.3  Cyclic Groups

Consider the group $D_4^+$. As we have seen above, we can write $D_4^+$ as $\{d^4, d, d^2, d^3\}$. Each element of this group can therefore be represented as a power of a single element, namely $d$. Such groups are called cyclic. More precisely:

**Definition 2.13** A group $G$ is called *cyclic* if it is generated by a single element.

Equivalent to this is: A group is called *cyclic* if there is a $g \in G$ such that all $h \in G$ can be written as $h = g^n$ for an integer $n$ (or, if the group operation is addition, as $h = g + \cdots + g$ or $h = -g - g \cdots - g$ with $n$ summands).

The groups $D_n^+$ are cyclic. They are generated by a rotation of $360/n$ degrees around the center of the regular $n$-gon. For example, the group $D_5^+$ is generated by the rotation corresponding to $(1, 2, 3, 4, 5)$.

```
gap> D5plus:=Group((1,2,3,4,5));
Group([ (1,2,3,4,5) ])
gap> Elements(D5plus);
[ (), (1,2,3,4,5), (1,3,5,2,4), (1,4,2,5,3), (1,5,4,3,2) ]
gap> Order(D5plus);
5
```

Also, the group $(\mathbb{Z}, +)$ is cyclic with 1 as a generator (0 is the identity element). Every positive integer can be represented as a sum of ones. What about the negative integers? The definition of a cyclic group also allows negative exponents of the generator. For example, $h = g^{-3}$ can also be written as $h = (g^{-1})^3$, i.e., take the inverse of $g$ to the power of three. As the group operation on the integers is addition, this means that we can write every integer as a sum where each term is either 1 or its inverse, i.e., $-1$. This gives us another characterization of cyclic groups:

A group is called *cyclic* if every element of the group can be written as a power of a single element or its inverse.

Is the group $D_3$ cyclic?

If you take all powers of a reflection and its inverse (which is the same reflection), you only get the identity and the reflection itself, because $s_a^2 = s_a \circ s_a = id$ and $s_a^3 = s_a$.

```
gap> G:=Group((1,2));
Group([ (1,2) ])
gap> Elements(G);
[ (), (1,2) ]
```

Each of the 3 reflections of $D_3$ thus only generates the identity and itself. The rotations $d_{120}$ or $d_{240}$ each only generate $\{id, d_{120}, d_{240}\}$. $D_3$ is therefore not cyclic. None of the elements of $D_3$ generates all elements.

Cyclic groups are commutative. Any two elements of a cyclic group can be written as $g^m$ and $g^n$ for integer $m$ and $n$. We then have $g^m g^n = g^{n+m} = g^n g^m$.

Let $\mathbb{Z}_n = \{0, 1, 2, \ldots, n-1\}$. Let $+_n$ be addition modulo $n$. For example, $3 +_4 2 = 1$, because multiples of 4 can be omitted (when dividing 5 by 4, the remainder is 1). Further examples are: $27 +_{35} 11 = 3$ and $6 +_8 4 = 2$. Be careful, some authors use the normal addition sign for addition modulo $n$. They write $3 + 2 \equiv 1 \bmod 4$.

**Theorem 2.14**  $(\mathbb{Z}_n, +_n)$ *is an abelian group.*

**Proof**  The identity element is 0. The inverse of $i$ is $n - i$. As with the addition of natural numbers, the associative law holds, and $+_n$ is commutative.                           □

Let's consider the group $D_5^+ = \{id, d, d^2, d^3, d^4\}$, the rotations in the regular pentagon with respect to composition. For example, $d^3 d^4 = d^2$, because if we rotate a regular pentagon first 4 times and then 3 times, it will end up in the same position as if we rotated it only 2 times instead.

In the group $\mathbb{Z}_5 = \{0, 1, 2, 3, 4\}$ we have $3 +_5 4 = 2$. Whether we calculate in the group $D_5^+$ or in $\mathbb{Z}_5$ makes no difference: A number modulo 5 is "the same" as the corresponding rotation in a regular pentagon. Adding two numbers modulo 5 is "the same" as the composition of 2 rotations in the regular pentagon.

Groups are referred to as "equal" or *isomorphic* when their group structure is the same. There must be a bijective map between both groups that transfers the group operation. We will clarify this term in Sect. 3.3. $\mathbb{Z}_5$ and $D_5^+$, or more generally, $\mathbb{Z}_n$ and $D_n^+$ are therefore isomorphic groups. More generally:

**Theorem 2.15**  *For a given natural number $n$, there is, up to isomorphism, only one cyclic group of order $n$.*

**Proof**  If we have a cyclic group $(G, \cdot)$ of order $n$ with generating element $g$ given, we rewrite the generating element as 1 and the identity element as 0. The element $g \cdot g$ we write as 2 etc., until we have assigned to each element of

$$G = \{id, g, g^2, \dots, g^{n-1}\}$$

an element of

$$\mathbb{Z}_n = \{0, 1, 2, \dots, n - 1\}.$$

Because $g^i \cdot g^j = g^{i+j \bmod n}$, the operation $\cdot$ in $G$ does exactly the same as the operation $+_n$ in $\mathbb{Z}_n$, because $i +_n j = i + j \bmod n$. Therefore, $G$ is isomorphic to $\mathbb{Z}_n$.                                                                                           □

We have written the group $D_3$ in two different ways:

$$D_3 = \{id, d_{120}, d_{240}, s_a, s_b, s_c\}$$

and

$$D_3' = \{(), (1, 2, 3), (1, 3, 2), (1, 2), (1, 3), (2, 3)\}.$$

**Fig. 2.3** Frieze pattern

In the first case, the group operation is the composition of isometries in the plane and in the second case, the composition of permutations. The elements of $D_3'$ describe the elements of $D_3$. The groups $D_3$ and $D_3'$ are isomorphic.

However, isomorphic groups can also come from completely different geometric figures. Consider the frieze pattern $F$ of Fig. 2.3. The symmetry group $G$ of $F$ contains only translations along the line $a$ by the vector $\vec{v}$ or its integer multiples. Let $t_v$ be the translation of $G$ along the vector $\vec{v}$. Then we can write $G$ as:

$$G = \{\ldots, t_v^{-3}, t_v^{-2}, t_v^{-1}, id, t_v, t_v^2, t_v^3, t_v^4, \ldots\}.$$

The symmetry group $G'$ of the frieze pattern $F'$ of Fig. 1.3 consists of glide reflections and their compositions. Let $\tau$ be the glide reflection which consists of the reflection over $a$ followed by the translation by the vector $\vec{v}$ in Fig. 1.3. Then we can write $G'$ as:

$$G' = \{\ldots, \tau^{-3}, \tau^{-2}, \tau^{-1}, id, \tau, \tau^2, \tau^3, \tau^4, \ldots\}.$$

$G$ and $G'$ are isomorphic, because the structure of the group $G$ is the same as that of the group $G'$ if you map $t_v$ to $\tau$. Here, $t_v^7 \circ t_v^2 = t_v^9$ just like $\tau^7 \circ \tau^2 = \tau^9$. Geometrically, however, something completely different happens: In $G$ there is a shift, and in $G'$ glide reflections are performed. So the geometric structure is not necessarily the same in isomorphic groups.

$G$ is also isomorphic to $(\mathbb{Z}, +)$ because we have the map $\phi \colon \mathbb{Z} \to G$ given by $i \to t_v^i$. If $i + j = k$, then $t_v^i \circ t_v^j = t_v^k$. The operation is therefore "the same". The number 3 corresponds to a shift of $F$ by 3 to the right, $-5$ corresponds to a shift by 5 to the left.

### Exercises

1. In the cyclic group $\mathbb{Z}_9$ calculate $2 +_9 3$ and $7 +_9 6$.
2. Is $G = \{0, 4, 8\}$ with addition modulo 12 a group?
3. Is $G = \{1, 3, 5, 7\}$ with multiplication modulo 8 a group?
4. Prove that the translations of a line in the direction of this line form a group. This group is isomorphic to the group $(\mathbb{R}, +)$, the real numbers with addition.
5. (a) Does $\mathbb{Z}_n$ with multiplication modulo $n$ form a group? (Hint: Consider what the identity element must be, and then check whether all elements have an inverse.)
   (b) Does $\mathbb{J}_n = \{1, 2, 3, \ldots, n - 1\}$ with multiplication modulo $n$ form a group? What property must the number $n$ have so that $\mathbb{J}_n$ forms a group?
6. Which elements of the cyclic group $\mathbb{Z}_{12}$ generate the group $\mathbb{Z}_{12}$ individually?

## 2.4   Properties of Groups

If $v, w, g$ are elements of a group, then $vgg^{-1}w = vw$, because $gg^{-1} = 1$, i.e. $vgg^{-1}w = v1w = vw$. The composition of an isometry followed by its inverse can just as well be omitted. Of course, this statement is not only valid for isometries. In every group $gg^{-1} = e$, where $e$ is the identity element of the group. For example:

```
gap> (1,4,2,6)(3,5)*((1,4,2,6)(3,5))^-1;
()
```

The inverse is denoted by `^-1` in GAP.

We define $g^0 = id$. We are forced to do this if we want the power law $g^n g^m = g^{n+m}$ to hold:

$$g^n = g^{n+0} = g^n g^0 = g^n 1 = g^n.$$

It holds that $(g^{-1})^{-1} = g$: If we want to undo the inverse of an isometry $g$, we execute $g$.

```
gap> ((1,2,5,3,6,4)^-1)^-1;
(1,2,5,3,6,4)
```

Earlier, we already used $g^{-n} = (g^{-1})^n$. This also follows from the above power law with

$$id = g^0 = g^{-n+n} = g^{-n}g^n = g^{-n}g \cdots g,$$

and each of the $g$ must be trivialized by a $g^{-1}$, so:

$$\underbrace{g^{-1} \cdots g^{-1}}_{n} = g^{-n}.$$

We summarize:

**Theorem 2.16** *Let $(G, \cdot)$ be a group and $v, w, g \in G$. Then the following hold:*

1. $v \cdot g \cdot g^{-1} \cdot w = v \cdot w$;
2. $g^0 = 1$;
3. $(g^{-1})^{-1} = g$;
4. $g^{-n} = (g^{-1})^n$.

There are other important elementary properties of groups:

**Theorem 2.17** *Let $(G, \cdot)$ be a group. Then the following hold:*

1. *In G there is only one identity element.*
2. *For each group element, there is exactly one inverse.*
3. *From $g \cdot v = g \cdot w$ or $v \cdot g = w \cdot g$ it follows that $v = w$ for group elements $g, v, w$.*

4. If $g_1, g_2, \ldots, g_n \in G$, then:

$$(g_1 \cdot g_2 \cdot \ldots \cdot g_n)^{-1} = g_n^{-1} \cdot g_{n-1}^{-1} \cdot \ldots \cdot g_1^{-1}.$$

## Proof

1. Let $e, e' \in G$ be identity elements, i.e., elements that fulfill (2.2) of Definition 2.1. Then $e \cdot e' = e$, since $e'$ is an identity element, and also $e \cdot e' = e'$, since $e$ is an identity element. It follows that $e = e'$.
2. Let $u, v \in G$ be inverses of $g \in G$. Then we have (where $e$ is the identity element in $G$):

$$u = e \cdot u = (v \cdot g) \cdot u = v \cdot (g \cdot u) = v \cdot e = v.$$

3. Multiply $g \cdot v = g \cdot w$ on both sides on the left by $g^{-1}$. This can be done, because if you have two identical group elements then they will remain equal if you multiply them each by the same element. Multiply $v \cdot g = w \cdot g$ correspondingly by $g^{-1}$ on the right.
4. $(g_1 \cdot g_2 \cdot \ldots \cdot g_n) \cdot (g_n^{-1} \cdot g_{n-1}^{-1} \cdot \ldots \cdot g_1^{-1}) = g_1 \cdot g_2 \cdot \ldots \cdot g_n \cdot g_n^{-1} \cdot g_{n-1}^{-1} \cdot \ldots \cdot g_1^{-1}$, and now cancel pairs of terms in the right-hand side of the equation from the middle (i.e., $g^n \cdot g^{-n} = e$, etc.), until the identity remains, i.e., $(g_n^{-1} \cdot g_{n-1}^{-1} \cdot \ldots \cdot g_1^{-1})$ is the inverse to $(g_1 \cdot g_2 \cdot \ldots \cdot g_n)$.

   You can also think of it this way: When you get dressed, you first put on a shirt and then a sweater. When undressing (undoing), you first take off the sweater and then the shirt.

   □

As a consequence of Theorem 2.17 item 3, we observe that in the multiplication table of a group (see Example 1.10) every group element occurs exactly once in each line and each column of the table. If there were two elements in a row which are the same, then there would be group elements $g, h, k$ with $gh = gk$ and $h \neq k$.

The cancellation rule 3 does not always hold when there is no group present. For example, when calculating modulo 12: $4 * 5 = 4 * 2$, because $4 * 5 = 20$ leaves the same remainder when divided by 12 as $4 * 2 = 8$. However, $5 \neq 2$ in $\mathbb{Z}_{12}$. There is no multiplicative inverse to 4 in $\mathbb{Z}_{12}$.

**Theorem 2.18** If $a, b, c, d$ are elements of a group $G$, the equations $xa = c$ and $by = d$ each have exactly one solution.

**Proof** To generate the isometry $c$ from the isometry $a$, first undo the isometry $a$ and then perform $c$, i.e., $ca^{-1}$. This sequence of operations is an isometry, which is exactly the element $x$. The theorem also holds for groups which are not symmetry groups: $ca^{-1}a = c$ is true in every group.

$by = d$ has the solution $b^{-1}d$ by a similar argument. □

**Exercises**

1. Let $a, b, c, d$ be elements of a group $G$. Assume the equation $ab^{-1}dca^{-1} = 1$
   holds in $G$ (1 is the identity element here). Solve this equation for $d$.
2. Solve the equation $(1, 3) \circ x = (1, 2)(3, 4)$ in the group $D_4$. (Hint: This is easy
   with GAP.)
3. What does $((g^{-1})^{-1})^{-1}$ yield for an element $g$ in a group $G$?


## 2.5  The Order of an Element

**Definition 2.19**  The *order* of an element $g$ in a group is the smallest number $n \in \mathbb{N}$
such that $g^n = e$ holds. One also writes $|g| = n$.

The natural numbers $\mathbb{N}$ do not contain 0. The smallest possible order is
therefore 1.

Every reflection has order 2 (Order gives the order     gap> Order((1,2));
of an element in GAP):                                      2

Elements of order 2 are called *involutions*.

In every group, the identity is the only element of     gap> Order(());
order 1:                                                       1

                                      gap> Order((1,3,5,7,9));

Further examples:                                        5

                                      gap> Order((1,4,8,5,3)(2,7,6));

                                      15

It is easy to see that the order of a permutation is equal to the least common
multiple of its cycle lengths.

In $\mathbb{Z}$ all elements, except the identity element, have infinite order.

The symmetry group of the frieze of Fig. 1.3 has infinite order, because it contains
infinitely many translations. Each of these translations (except the identity) has
infinite order.

The order of the translations of the symmetry group $G_{(4,4)}$ of the decomposition
of the plane of Example 1.2 is infinite. A translation always has infinite order: If
you perform a shift of the plane several times in succession, the identity can never
result.

The order of a rotation by $360/n$ degrees is $n$ for each $n \in \mathbb{N}$. For other angles $\alpha$,
the order of the corresponding rotation is the smallest number $k \in \mathbb{N}$ such that $k\alpha$ is
a multiple of $360°$ (if such a $k$ exists).

If two groups are isomorphic, they have the same number of elements. The
groups $(\mathbb{Z}_4, +_4)$ and $(D_2, \circ)$ each have four elements. Are they isomorphic? In $\mathbb{Z}_4$,
the number 1 has order 4 (i.e. $|1| = 4$). However, in the group $D_2$ there is no element
of order 4: $D_2$ is the symmetry group of a "regular 2-gon", i.e., a line segment (see
Fig. 2.4). It contains two reflections $s_a$ and $s_b$ of order 2 and a rotation by $180°$,
which also has order 2. Therefore, the groups $D_2$ and $\mathbb{Z}_4$ are not isomorphic. The
reflection $s_a$ in Fig. 2.4 is an isometry of the line segment, because it is an isometry

**Fig. 2.4** Line segment

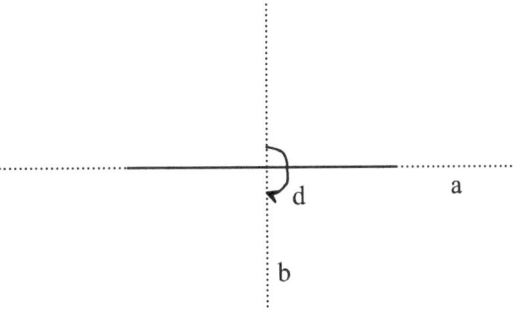

of the plane, which maps the line segment onto itself. It is therefore not equal to the identity.

**Definition 2.20**  The group $D_2 = \{id, d, s_a, s_b\}$ is called *Klein's four-group*, after the mathematician FELIX KLEIN (1849–1925).

The group of the rhombus of Example 2.8 on page 18 is isomorphic to $D_2$, because it consists of the same isometries (see Fig. 1.5): Besides the identity, these are two reflections along perpendicular mirror axes and a rotation about their intersection point by 180°. Any figure in the plane with these symmetries has the same group, including a rectangle, which is not a square.

**Theorem 2.21**  *If the order of an element $g \in G$ is finite and equal to the order of the group $G$, then $G$ is cyclic and is generated by $g$.*

*Proof*  If we can show that $G$ is generated by $g$, then we will have proven that $G$ is cyclic. We therefore only need to show that $g$ generates the group $G$.

Let $|g| = |G| = n$. So $g^n = id$, but $g^i \neq id$, $\forall i < n$. The set $\{id, g, g^2, \ldots, g^{n-1}\}$ consists of mutually different elements: It follows from $g^i = g^j$ by Theorem 2.17 item 3 (see page 24) that $g^{i-j} = 1$ and this is only true for $i = j$. Because $|G| = n$ it follows that

$$G = \{id, g, g^2, \ldots, g^{n-1}\}.$$

Every element of the group can therefore be written as a $g$-power. Therefore $g$ generates the group $G$.                                                                    □

In the last example of this chapter, we consider isometries in $\mathbb{R}^3$.

**Example 2.22**  The symmetry group $S_4$ of the tetrahedron, the *tetrahedral group*, has order 24.

Consider the tetrahedron of Fig. 2.5. Here, the permutation notation is very helpful. Each isometry of $\mathbb{R}^3$ which maps this tetrahedron onto itself can be uniquely described by the images of the vertices of the tetrahedron and thus by a permutation of the numbers 1, 2, 3, 4.

**Fig. 2.5** Tetrahedron

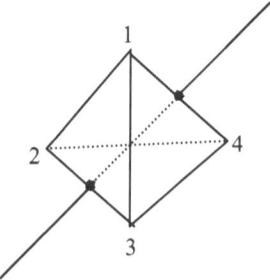

Which rotations does the tetrahedron allow? For example, there are rotations about the axis through point 1 and the center of the opposite triangular face. These rotations can be written as $(2, 3, 4)$, $(2, 3, 4)^2 = (2, 4, 3)$, $(2, 3, 4)^3 = id$. The permutation $(2, 3, 4)$ thus describes a rotation of order 3. There are 4 rotation axes of this type, each passing through one of the vertices of the tetrahedron and the center of the opposite face. So far, we have 9 elements of the group $S_4$:

$$S_{4,1} = \{id, (2, 3, 4), (2, 4, 3), (1, 3, 4), (1, 4, 3), (1, 2, 4),$$

$$(1, 4, 2), (1, 2, 3), (1, 3, 2)\}.$$

We can perform a 180° rotation around the axis that runs through the midpoint of edge 1, 4 and the midpoint of edge 2, 3 (see Fig. 2.5). This rotation is written as $(2, 3)(1, 4)$. It has order 2. There are three rotations of this type, one for each pair of opposite edges:

$$S_{4,2} = \{(2, 3)(1, 4), (1, 2)(3, 4), (1, 3)(2, 4)\}.$$

There are also orientation-reversing isometries. One can reflect, for example, through the plane that runs through the vertices 1 and 3 and the midpoint of the edge 2, 4. This reflection corresponds to the permutation $(2, 4)$. Like every reflection, it has order 2. For each of the six edges of the tetrahedron, there is such a reflection. Here are all these reflections:

$$S_{4,3} = \{(2, 4), (1, 2), (1, 3), (1, 4), (2, 3), (3, 4)\}.$$

The remaining six isometries are harder to visualize. But we can write them as a composition of a plane reflection from $S_{4,3}$ with a rotation from $S_{4,1}$. For example: $(2, 4)(1, 2, 3) = (1, 4, 2, 3)$. There are six isometries of this type:

$$S_{4,4} = \{(1, 2, 4, 3), (1, 2, 3, 4), (1, 3, 2, 4), (1, 3, 4, 2), (1, 4, 2, 3), (1, 4, 3, 2)\}.$$

In total, we have thus counted the claimed 24 isometries. There can't be more, because there are only $24 = 4 * 3 * 2 * 1$ permutations of the numbers $1, 2, 3, 4$. All permutations of 4 numbers thus form a group, the *symmetric group* $S_4$, and this is isomorphic to the tetrahedral group. That's why we called the tetrahedral group $S_4$.

More generally, the group of all permutations of $n$ elements with respect to composition is called the *symmetric group* $S_n$. Details can be found in Sect. 4.1.

The tetrahedral group can be defined in GAP. To do this, we observe that we can obtain all elements of the tetrahedral group by composition of elements of $S_{4,3}$ (see Exercise 5.). Even fewer elements are sufficient; which ones?

```
gap> Tetra:=Group((2,4),(1,2),(1,3),(1,4),(2,3),(3,4));
Group([ (2,4), (1,2), (1,3), (1,4), (2,3), (3,4) ])
gap> Order(Tetra);
24
gap> Elements(Tetra);
[ (), (3,4), (2,3), (2,3,4), (2,4,3), (2,4), (1,2),
  (1,2)(3,4), (1,2,3), (1,2,3,4), (1,2,4,3), (1,2,4),
  (1,3,2), (1,3,4,2), (1,3), (1,3,4), (1,3)(2,4),
  (1,3,2,4), (1,4,3,2), (1,4,2), (1,4,3), (1,4),
  (1,4,2,3), (1,4)(2,3) ]
```

The tetrahedral group is thus generated by the elements of $S_{4,3}$.

In principle, a group can also require infinitely many generators. Every finitely generated group has only countably many group elements: We first count the identity, then all group elements of length 1 (i.e., those that can be represented by a generator or the inverse of a generator), then all group elements of length 2, etc. Since there are only finitely many group elements for a given length, this procedure leads to a count of all group elements. Since the real numbers are uncountable, we have proven:

**Theorem 2.23** *The real numbers with ordinary addition form a non-finitely generated group.*

However, we will consider, with few exceptions, only *finitely generated* groups. Here comes another exception:

**Theorem 2.24** *The isometry group $\mathcal{E}$ of the Euclidean plane is generated by all reflections.*

**Proof** According to Theorem 1.9, we obtain every translation as the product of two reflections (along parallel lines perpendicular to the direction of translation with a distance equal to half the length of the translation vector). The same theorem proves every rotation is the product of two reflections (the two corresponding lines intersect at the rotation point with half the rotation angle). Theorem 1.6 on page 9 shows that every isometry of the plane is either a translation, reflection, glide reflection, or rotation. We only need to show that we can generate a glide reflection through reflections.

A glide reflection is a reflection followed by a translation and can therefore be generated by three reflections, two of which are for the translation.                □

$\mathcal{E}$ is of course not finitely generated. Even the symmetry group of a circle in the plane is, according to Exercise 6., not finitely generated.

**Exercises**

1. Determine the order of 3 in $(\mathbb{Z}_7, +_7)$ and of 5 in $(\mathbb{Z}_{20}, +_{20})$. (Hint: Use GAP.)
2. What order does $1/3$ have in $(\mathbb{Q} - \{0\}, *)$?
3. Specify a group that contains elements of order 2, 3, and 4. (Hint: Reflections always have order 2. Are there rotations of order 3 and 4 in a regular $n$-gon?)
4. Prove that a group is abelian if each of its elements has order 2.
5. Generate the elements of $S_{4,2}$ (see Example 2.22) from the elements of $S_{4,3}$. Generate the elements of $S_{4,1}$ from the elements of $S_{4,3}$. If you have trouble composing the elements, use GAP.
6. Show that the symmetry group of a circle is infinitely generated.
7. What order does the symmetry group of an octahedron have?
8. Generate with GAP a multiplication table with the command

$$\texttt{MultiplicationTable(Elements(G));}$$

where G must be a finite (preferably small) group defined in GAP (also read the corresponding topic in the GAP documentation). How can you see, just by looking at the table, whether the group is abelian or not?
9. Show that the group of rational numbers with respect to addition is infinitely generated.
10. Let $G$ be a group and $g \in G$. Prove that $g$ and $g^{-1}$ have the same order. (Hint: Use Theorem 2.16 item 4).
11. Let $G$ be a group and $g, h \in G$ such that $g$ and $h$ commute, that is $gh = hg$. Show $|gh| = |g| \cdot |h|$.

# Chapter 3
# Subgroups and Homomorphisms

In this chapter, we examine subgroups, i.e., subsets of groups that form groups themselves, and maps (homomorphisms) between groups that transfer the group structure. We gain initial insights into which subgroups can occur in groups (Lagrange's theorem), and we precisely define when two groups are to be considered "equal" (isomorphic). In addition, we assign a subgroup with certain properties (a normal subgroup) to each homomorphism. In the last section, we apply our findings to the subgroup of translations of the symmetry group of the plane.

## 3.1 Subgroups

**Example 3.1** Consider the symmetry group $D_6$ of the regular hexagon of Fig. 3.1. Let $U$ be the set of isometries that transforms the set of points $\{1, 3, 5\}$ into itself.

$U$ contains three reflections. The corresponding reflection axes are drawn in. In addition, $U$ contains the identity and rotations of 120 and 240°. $U$ is the group of the regular triangle. In fact, $U$ consists of exactly the isometries that transform the triangle with vertices 1, 3, and 5 onto itself. At the same time, $U$ is a subset of $D_6$. The group $U$ is the *stabilizer* of the set of points $\{1, 3, 5\}$ (i.e., each element of $U$ transforms this set of points onto itself) and it is a *subgroup* of $D_6$ (i.e. a subset of $D_6$ that forms a group itself).

In Sect. 2.2 we defined the set $D_n^+$ as the set of rotations of $D_n$ including the element $id$ (which is a rotation by 0°). As we have already seen, we have

$$D_n^+ = \{id, d, d^2, d^3, \ldots, d^{n-1}\}.$$

We also realized that these rotations form a group with respect to composition. The set $D_n^+$ is a subset of $D_n$ that forms a group with the same operation as $D_n$.

S. Rosebrock, *Visual Group Theory*, Springer Undergraduate Mathematics Series, https://doi.org/10.1007/978-3-662-69365-0_3

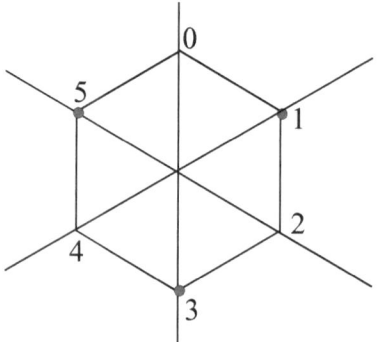

**Definition 3.2** A subset $U$ of a group $G$ is called a *subgroup* of $G$ if $U$ forms a group itself with the operation of $G$.

We write $U < G$ if $U$ is a subgroup of $G$. So $D_n^+ < D_n$. Every group contains itself as a subgroup, so $G < G$. In addition, every group contains the *trivial group* as a subgroup. This is the group that consists only of the identity element. We often denote it by $\{e\}$. A subgroup of a group $G$ which is neither the trivial group nor $G$ itself is called a *proper subgroup* of $G$.

To look at subgroups in GAP, we first define a group. In Exercise 5. of Sect. 2.1 it was proven that the group $D_n$ is generated by a reflection $s \in D_n$ and a rotation around the center of the regular $n$-gon by $360/n$ degrees. So we can define the group $D_4$ in GAP by the reflection over the line through vertices 2 and 4 (described by the permutation $(1,3)$) and the clockwise rotation by $90°$ around the center (described by the permutation $(1,2,3,4)$). See Fig. 1.6. The command Subgroup generates a subgroup. Here too, we only need to specify the generators of the subgroup. We want to generate $D_4^+$. This group is cyclic, so it is generated by only one element, a rotation by $90°$:

```
gap> D4:=Group((1,3),(1,2,3,4));
Group([ (1,3), (1,2,3,4) ])
gap> D4plus:=Subgroup(D4,[(1,2,3,4)]);
Group([ (1,2,3,4) ])
gap> Elements(D4plus);
[ (), (1,2,3,4), (1,3)(2,4), (1,4,3,2) ]
```

The group $G_{(4,4)}$ of the decomposition of the plane into squares of Example 2.6 on page 16 has the group $D_4$ of the square as a subgroup. This subgroup is generated by the elements $s_a, s_b$ (see Fig. 2.1). It is the stabilizer of the intersection of the axes $a$ and $b$ in $G_{(4,4)}$.

The group of the square appears as a subgroup infinitely often: In each of the squares of the decomposition of Fig. 1.4, two mirror axes analogous to the axes $a, b$ can be laid. The corresponding reflections then generate another subgroup isomorphic to $D_4$. The stabilizer of the center of an arbitrary square is thus a subgroup $D_4$. But also the stabilizer of the intersection of a horizontal and a vertical line of Fig. 1.4 forms a subgroup that is isomorphic to $D_4$.

In the first chapter, we considered orientation-preserving and orientation-reversing isometries. Orientation-reversing isometries of the plane are reflections and glide reflections, where the plane is "flipped" (see Theorem 1.6).

Orientation-preserving isometries are rotations and translations. If you compose two orientation-preserving isometries, the result is again an orientation-preserving isometry. The inverse of an orientation-preserving isometry is again orientation-preserving. The identity map is orientation-preserving. All orientation-preserving isometries of the plane thus form a group and therefore, according to Definition 3.2:

**Theorem 3.3** *The orientation-preserving isometries $\mathcal{E}^+$ of the plane form a subgroup of the group $\mathcal{E}$ of the isometries of the plane.*

By the same argument one sees:

**Theorem 3.4** *The orientation-preserving isometries of a figure $F$ in the plane form a subgroup of the symmetry group of $F$.*

If $G$ is the symmetry group of a figure $F$, we denote the subgroup of orientation-preserving isometries by $G^+$. This explains the notation $D_n^+$. The rotations are just the orientation-preserving isometries of the regular $n$-gon.

Let $F$ be any figure in the plane (or a solid in $\mathbb{R}^3$) and $S \subset F$. Let $G$ be the symmetry group of $F$. Then the elements of $G$ that map $S$ onto $S$ form a subgroup $G(S) < G$, the *stabilizer* of $S$. For $u \in G(S)$, the map $u^{-1}$, which undoes $u$, also maps $S$ onto $S$, so $u^{-1} \in G(S)$. If $u, v \in G(S)$ then $u \circ v \in G(S)$. In Example 3.1 the group $D_3$ is the stabilizer of the vertices 1,3,5 in the regular hexagon, and therefore $D_3 < D_6$.

**Definition 3.5** $O_2 < \mathcal{E}$ is the subgroup whose elements fix the origin, the *orthogonal group*.

The orthogonal group is thus the stabilizer of the origin in the symmetry group of the plane.

The stabilizer of the vertex 1 of Example 2.22 on page 27 is the subgroup consisting of all isometries that fix the vertex 1. The vertices 2,3,4 can therefore be arbitrarily permuted. Therefore, for the stabilizer $S_4(1) = D_3$ (last is the last result in GAP):

```
gap> Tetra:=Group((2,4),(1,2),(1,3),(1,4),(2,3),(3,4));
Group([ (2,4), (1,2), (1,3), (1,4), (2,3), (3,4) ])
gap> Stabilizer(Tetra,1);
Group([ (3,4), (2,3,4) ])
gap> Elements(last);
[ (), (3,4), (2,3), (2,3,4), (2,4,3), (2,4) ]
```

We obtain the stabilizer of the edge 1,2 of the tetrahedron as follows:

```
gap> Stabilizer(Tetra,[1,2],OnSets);
Group([ (1,2)(3,4), (3,4) ])
gap> Elements(last);
[ (), (3,4), (1,2), (1,2)(3,4) ]
```

We can give many more examples of
subgroups: Every reflection of a figure
generates a subgroup of order 2 in the
symmetry group of the figure:

```
gap> Subgroup(D4,[(1,3)]);
Group([ (1,3) ])
gap> Elements(last);
[ (), (1,3) ]
```

If you compose two translations of the plane, you get another translation. For each translation, the inverse map is again a translation. The identity can be considered as a translation by the 0-vector. Translations preserve orientation. Therefore:

**Theorem 3.6** *The translations form a subgroup of $\mathcal{E}^+$.*

This subgroup of translations is called $\mathcal{T}$.

**Theorem 3.7** $(\mathbb{Z}, +) < (\mathbb{Q}, +)$.

**Proof** The rational number 0 is an integer. The identity element of $(\mathbb{Q}, +)$ is thus also the identity element of $(\mathbb{Z}, +)$. The inverse of an integer is an integer. □

We consider the group $(\mathbb{Z}_8, +_8)$ with $\mathbb{Z}_8 = \{0, 1, 2, 3, 4, 5, 6, 7\}$. The group $(U, +_8)$ with $U = \{0, 2, 4, 6\}$ forms a subgroup of the group $\mathbb{Z}_8$, because: For the identity element 0 we have $0 \in U$, and the sum (mod 8) of two even numbers is again even (e.g. $4 +_8 6 = 2$). The inverses are also present (e.g. $2 +_8 6 = 0$ or $4 +_8 4 = 0$).

The set $H = \{0, 2, 4, 5\}$ does not form a subgroup of $\mathbb{Z}_8$, because for example $2 +_8 5 = 7 \notin H$.

**Example 3.8** A rectangle can be inscribed into a regular octagon, as shown in Fig. 3.2. Every isometry of the rectangle is also an isometry of the regular octagon. Thus, the group of the rectangle (the Klein four-group: see Example 2.20 on page 27) is a subgroup of the group $D_8$. It is the stabilizer of the vertex set $\{1, 4, 5, 8\}$.

In GAP with the vertex labels of Fig. 3.2 we generate the group $D_8$ with a reflection and a clockwise rotation of $45°$.

```
gap> D8:=Group((1,2)(8,3)(7,4)(6,5),(1,2,3,4,5,6,7,8));
Group([ (1,2)(3,8)(4,7)(5,6), (1,2,3,4,5,6,7,8) ])
gap> D2:=Subgroup(D8,[(1,4)(8,5)(2,3)(7,6),
                      (1,8)(4,5)(2,7)(3,6)]);
Group([ (1,4)(2,3)(5,8)(6,7), (1,8)(2,7)(3,6)(4,5) ])
gap> Elements(D2);
[ (), (1,4)(2,3)(5,8)(6,7), (1,5)(2,6)(3,7)(4,8),
      (1,8)(2,7)(3,6)(4,5) ]
```

Here, $(1,2)(8,3)(7,4)(6,5)$ corresponds to the reflection over the line through the midpoints of edges 1, 2 and 5, 6. The element $(1,4)(8,5)(2,3)(7,6)$ corresponds to the reflection over the vertical line through the rectangle's center and $(1,8)(4,5)(2,7)(3,6)$ over the horizontal line. To define the symmetry group of the rectangle, four vertices would suffice in GAP, but as a subgroup of the group $D_8$, all eight vertices of the octagon are necessary.

**Fig. 3.2**  A rectangle in a
regular octagon

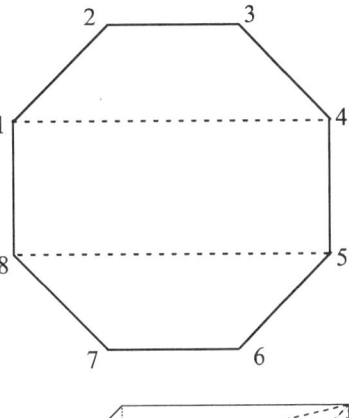

**Fig. 3.3**  Tetrahedron in the
cube

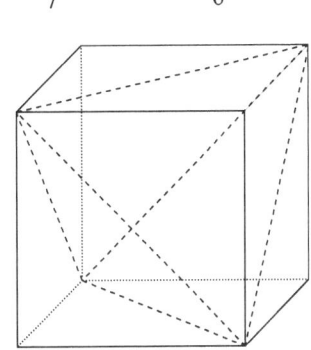

The tetrahedral group $S_4$ of Example 2.22 on page 27 is a subgroup of the cube group (see Sect. 1.4). Indeed, the tetrahedron can be inscribed in the cube, as shown in Fig. 3.3, and we can see that every isometry of the tetrahedron is also an isometry of the cube. The group $S_4$ thus appears as the stabilizer of 4 selected vertices of the cube in the cube group.

The following is a very general criterion for subgroups:

**Theorem 3.9**  *A non-empty subset H of a group G is a subgroup of G if and only if*

$$\forall a, b \in H \qquad ab^{-1} \in H.$$

***Proof***  We show that if the above condition is fulfilled then $H$ is a subgroup: $H$ is not empty, so there is a $g \in H$. Associativity holds in $H$, because its elements are in $G$ and associativity holds in $G$.

Existence of the identity element: With $g, g \in H$ inserted into the above condition, it follows that $gg^{-1} = e \in H$.

Existence of the inverse: If $a \in H$ then (insert $e, a$ into the above condition) $ea^{-1} = a^{-1} \in H$.

Closure: If $a, b^{-1} \in H$ then $a(b^{-1})^{-1} = ab \in H$. So $H$ is a subgroup of $G$.

The converse is clear.                                                                      □

But one can also pick some elements of a group and ask which subgroup these elements generate: In the symmetry group of the regular octagon $D_8$ we consider the subgroup $U$ which is generated by a rotation by $90°$, so $U = \langle d_{90} \rangle$. We have to add composites of elements to $\{d_{90}\}$ until the resulting set forms a subgroup. Since $d_{90} \in U$, also $d_{90} \circ d_{90}$ and $d_{90} \circ d_{90} \circ d_{90}$ must be in $U$. Every subgroup must contain the identity. So it follows that

$$U = \{id, d_{90}, d_{90} \circ d_{90}, d_{90} \circ d_{90} \circ d_{90}\}$$

or, written more simply,

$$U = \{id, d_{90}, d_{90}^2, d_{90}^3\}.$$

Thus, the set $U$ is closed with respect to inverses and all products are included. Therefore, $U$ is a subset of $D_8$ which itself is a group, so it is a subgroup of $D_8$.

We generalize to more generators: Let $G$ be a group and $g_1, g_2, \ldots, g_n \in G$. Then

$$U = \langle g_1, g_2, \ldots, g_n \rangle$$

is the smallest subgroup of $G$ which contains the elements $g_1, g_2, \ldots, g_n$. Such a subgroup always exists, and it may be the whole group $G$. The subgroup $U$ is called *the subgroup generated by* $g_1, g_2, \ldots, g_n$. If $n = 1$ (so $U = \langle g_1 \rangle$), then the resulting subgroup is cyclic, because it is generated by the sole element $g_1$.

As another example, we consider the group $(\mathbb{Z}, +)$ with the subgroup $U = \langle 2 \rangle$. Because $2 \in U$, $-2 \in U$ and $2 + 2 = 4 \in U$ must also be in $U$. With further arguments of the same type, it follows that

$$U = \{\ldots, -6, -4, -2, 0, 2, 4, 6, 8, \ldots\}.$$

The set of all even numbers thus forms a subgroup of $(\mathbb{Z}, +)$. This is also easy to see in another way: The sum of two even numbers is even, 0 (the identity element) is even and the negative of an even number is even.

**Theorem 3.10** *A subgroup of a cyclic group is cyclic.*

**Proof** Let $G = \langle g \rangle$ be a cyclic group and $U < G$ a nontrivial subgroup (the trivial group is by definition cyclic). Let $g^n \in U$ be chosen such that the exponent $n$ has the minimal possible absolute value among all powers of $g$ not equal to the identity. Let $g^k \in U$ be any element of the subgroup. It is now possible to perform the division with remainder of $k$ by $n$, i.e.

$$k = ni + r, \quad 0 \le r < |n|.$$

So $g^k = g^{ni} \circ g^r$ or, in other words, $g^r = g^k \circ (g^n)^{-i}$ (after multiplying both sides by $g^{-ni}$). The right-hand side is an element of $U$ and therefore $g^r \in U$. But since $|n|$

is minimal, it follows that $r = 0$ and therefore $g^k = (g^n)^i$. So every element of the subgroup can be expressed as a power of $g^n$, and therefore $U = \langle g^n \rangle$ is cyclic.   □

**Exercises**

1. Prove that $(\mathbb{Q} - \{0\}, *) < (\mathbb{R} - \{0\}, *)$.
2. Prove that the intersection of two subgroups of a group $G$ forms a subgroup of $G$.
3. Do all rotations form a subgroup of the isometry group of the plane?
4. Let $G = \langle a, b \rangle$ be a group generated by two elements. Show that if $ab = ba$ in $G$, then $G$ is abelian.
5. Prove that the elements of finite order in an abelian group form a subgroup.
6. Determine all cyclic groups.
7. Prove that for $n \in \mathbb{N}$ there are only finitely many groups of order $n$. Can you give an upper bound for the number of these groups?
8. What orders can the elements of $\mathbb{Z}_{18}$ have? Name an element for each possible order.

   Hint: If you have theoretical difficulties, use the GAP command
   ```
   Z18:=Group((1,2,3,4,5,6,7,8,9,10,11,12,13,14,15,16,
   17,18));
   ```
9. Determine all groups of orders 1,2,3,4 and 5. In particular, prove that the only groups of order 4 are the group $\mathbb{Z}_4$ and the Klein four-group.

## 3.2   Cosets and Lagrange's Theorem

We consider the rotation group $D_n^+ < D_n$ as a subgroup of the symmetry group of the regular $n$-gon. Let $s_1, \ldots, s_n \in D_n$ be the reflections. If we combine a random reflection with $s_1$, we get a rotation: $s_1 \circ s_i = d$, or, after multiplication by $s_1$ on the left, $s_i = s_1 \circ d$. If we multiply all reflections of the group $D_n$ by $s_1$ in this way, we get all rotations, or in other words, we get all reflections by multiplying $s_1$ by all rotations, so:

$$D_n = \{id, d, d^2, \ldots, d^{n-1}, s_1, s_1 d, s_1 d^2, \ldots, s_1 d^{n-1}\}.$$

We write $s_1 D_n^+$ for $\{s_1, s_1 d, s_1 d^2, \ldots, s_1 d^{n-1}\}$. So we have: $D_n = D_n^+ \cup s_1 D_n^+$.

In general: Let $H$ be a subgroup of the group $G$ and $g \in G$. Then $gH = \{gh \mid h \in H\}$ is a *left coset* of $G$. The elements of $gH$ form a subset of $G$.

**Example 3.11** In GAP we consider the pair $D_4^+ < D_4$ and the left coset $(2, 4) D_4^+$:

```
gap> D4:=Group((1,3),(1,2,3,4));
Group([ (1,3), (1,2,3,4) ])
gap> D4plus:=Subgroup(D4,[(1,2,3,4)]);
Group([ (1,2,3,4) ])
gap> Elements(D4plus)*(2,4);
```

```
[ (2,4), (1,2)(3,4), (1,3), (1,4)(2,3) ]
```

There are corresponding *right cosets*: $Hg = \{hg \mid h \in H\}$. Remember that in GAP elements are concatenated from left to right. We concatenate from right to left.

```
gap> (2,4)*Elements(D4plus);
[ (2,4), (1,4)(2,3), (1,3), (1,2)(3,4) ]
```

In this particular case, the left and right cosets are equal. This does not always have to be the case, and we will later see examples where $gH \neq Hg$ (see Example 3.19).

**Theorem 3.12** *Let G be a group and $H < G$ a subgroup. Then G can be written as a disjoint union of left cosets.*

**Proof** We have

$$G = \bigcup_{g \in G} gH \tag{3.1}$$

($G$ is the union of the left cosets over all elements $g \in G$), because the identity element $e \in H$. The coset $gH$ therefore at least contains the element $g$.

We show that if two left cosets $aH$ and $bH$ have an element $c$ in common, then they are equal. So let $c = ah_1$ and $c = bh_2$, where $h_1, h_2 \in H$. Then $a = ch_1^{-1} = bh_2h_1^{-1}$. Every element $ah \in aH$ thus has the form $ah = b(h_2h_1^{-1}h) \in bH$. Therefore, $aH \subset bH$. Similarly, one shows $bH \subset aH$, so that $aH = bH$ holds. Now delete in (3.1) every coset that appears more than once. Then the union is disjoint.                                                                                     □

The theorem naturally applies to right cosets as well and is proven analogously. In particular, from the proof of the theorem we have the following.

**Lemma 3.13** *If G is a group and $H < G$, then $\forall a, b \in G$:*

$$b \in aH \Rightarrow aH = bH.$$

The following theorem of Lagrange was actually only proven by him for groups of permutations. Groups in today's form were not yet known at that time.

**Theorem 3.14 (Lagrange 1771)** *The order of a subgroup H of a finite group G is a divisor of the order of G.*

**Proof** We show that (for finite groups) $|aH| = |H|, \forall a \in G$. Then Theorem 3.12 proves that the number of elements in $H$ multiplied by the number of cosets gives the order of $G$.

We even show that there is a bijective map $\phi_a : H \to aH$ defined by $\phi_a(h) = ah$, which is the *left multiplication by a*. The injectivity can be seen as follows: From $ah_1 = ah_2$ it follows that $h_1 = h_2$. Moreover, $aH$ has at most as many elements as $H$, and from this it follows that $\phi_a$ must be a bijection.                                           □

**Definition 3.15** Let $G$ be a group and $H < G$ a subgroup. The *index* of $H$ in $G$ is the number of cosets of $H$ in $G$. We also write $[G : H]$ for this number.

In GAP:
```
gap> Index(D4, D4plus);
2
```
The index here is 2, because the group $D_4^+$ has exactly half as many elements as the group $D_4$.

In the proof of Theorem 3.14 we saw that each coset has the same number of elements. This implies the following

**Corollary 3.16** *If $G$ is a finite group and $H$ is a subgroup of $G$, then $|G| = |H| \cdot [G : H]$.*

It is now easy to prove the product theorem for the index.

**Theorem 3.17** *If $G$ is a finite group and $J < H < G$, then*

$$[G : J] = [G : H] \cdot [H : J].$$

**Proof** $[G : J] = |G|/|J| = (|G|/|H|) \cdot (|H|/|J|) = [G : H] \cdot [H : J].$ □

**Example 3.18** In Example 2.6 on page 16 the group $G_{(4,4)}$ is given as the symmetry group of the decomposition of the plane into squares. As explained in Sect. 3.1, there is a subgroup $U < G_{(4,4)}$ isomorphic to the group $D_4$ of the square generated by the elements $s_a, s_b$ of Fig. 2.1. The index $[G_{(4,4)} : U]$ is infinite, because $G_{(4,4)}$ is infinite and $U$ is finite.

**Example 3.19** We consider once again Example 3.8 on page 34:
```
gap> D8:=Group((1,2)(8,3)(7,4)(6,5),(1,2,3,4,5,6,7,8));
Group([ (1,2)(3,8)(4,7)(5,6), (1,2,3,4,5,6,7,8) ])
gap> D2:=Subgroup(D8,[(1,4)(8,5)(2,3)(7,6),
                      (1,8)(4,5)(2,7)(3,6)]);
Group([ (1,4)(2,3)(5,8)(6,7), (1,8)(2,7)(3,6)(4,5) ])
gap> Index(D8, D2);
4
```
So there are 4 times as many elements in the group $D_8$ as in the group $D_2$, or in other words, there are 4 left cosets of $D_2$ in $D_8$: One coset we get through the elements of $D_2$ itself:
```
gap> Elements(D2);
[ (), (1,4)(2,3)(5,8)(6,7), (1,5)(2,6)(3,7)(4,8),
  (1,8)(2,7)(3,6)(4,5) ]
```
Take an arbitrary element that is not in $D_2$, and generate another coset with it:
```
gap> (1,2,3,4,5,6,7,8)*Elements(D2);
[ (1,2,3,4,5,6,7,8), (1,3)(4,8)(5,7), (1,6,3,8,5,2,7,4),
  (1,7)(2,6)(3,5) ]
```
For practice, you should trace these isometries in Fig. 3.2. Again, we take an element that does not occur in the two cosets we have already generated, and generate another coset with it:

```
gap> (1,3,5,7)(2,4,6,8)*Elements(D2);
[ (1,3,5,7)(2,4,6,8), (1,2)(3,8)(4,7)(5,6),
   (1,7,5,3)(2,8,6,4), (1,6)(2,5)(3,4)(7,8) ]
```

and

```
gap> (1,4,7,2,5,8,3,6)*Elements(D2);
[ (1,4,7,2,5,8,3,6), (2,8)(3,7)(4,6),
   (1,8,7,6,5,4,3,2), (1,5)(2,4)(6,8) ]
```

Convince yourself that we have listed all elements of the group $D_8$ in these 4 left cosets.

In this case, by the way, the left cosets are not equal to the right cosets, because:

```
gap> Elements(D2)*(1,4,7,2,5,8,3,6);
[ (1,4,7,2,5,8,3,6), (1,7)(2,6)(3,5),
   (1,8,7,6,5,4,3,2), (1,3)(4,8)(5,7) ]
```

and this coset is different from the above coset

```
(1,4,7,2,5,8,3,6)*Elements(D2).
```

Let $G$ be a group. Choose $g \in G$ with $g \neq e$. The elements

$$\langle g \rangle = \{\ldots, g^{-2}, g^{-1}, id, g, g^2, \ldots\}$$

form a cyclic subgroup $H < G$, the subgroup generated by $g$. The subgroup $H$ can indeed be finite, namely when repetitions occur in the sequence

$$\ldots, g^{-2}, g^{-1}, id, g, g^2, \ldots.$$

For example, the subgroup $\langle 2 \rangle$ in $\mathbb{Z}_8$ has order 4.

**Corollary 3.20** *Every group whose order is a prime number is cyclic.*

**Proof** Let $G$ be a group of prime order. Choose $g \in G$ with $g \neq id$. Let $H$ be the subgroup generated by $g$. The order of $G$ is finite, and Lagrange's theorem says that the order of $H$ divides that of $G$. Since $|G|$ is prime, it follows that $|H| = 1$ or $|H| = |G|$. Obviously, $|H|$ is greater than 1, because $id, g \in H$. So $G \cong H$, and $g$ is a generator of the cyclic group $G$.                                                                      □

**Corollary 3.21** *Let $G$ be a finite group of order $n$ and $g \in G$. Then the order of $g$ is a divisor of $n$ and $g^n = id$.*

**Proof** Because $G$ is a finite group, there are repetitions in the sequence

$$id, g, g^2, g^3, \ldots.$$

So there exist $i < j$ such that $g^j = g^i$, and $p = j - i$ is minimal. Multiplying on both sides by $g^{-i}$, we get $g^p = id$. Because $p$ was chosen minimally, $g$ has order $p$. The subgroup generated by $g$ is therefore finite cyclic of order $p$. According to

Lagrange's theorem, $p$ divides the order of $G$, which shows the first part of the assertion. So there is a $k \in \mathbb{N}$ with $n = p \cdot k$, i.e. $g^n = g^{pk} = id^k = id$. $\qquad \square$

**Exercises**

1. Show, by similar arguments as those of Example 3.1 on page 32, that $D_n < D_{2n}$. What is $[D_{2n} : D_n]$?
2. Show that $6\mathbb{Z} = \{6 \cdot k \mid \forall k \in \mathbb{Z}\}$ is a subgroup of $(\mathbb{Z}, +)$. More generally, show that for all natural numbers $n \geq 2$, $n\mathbb{Z} = \{n \cdot k \mid \forall k \in \mathbb{Z}\}$ is a subgroup of $(\mathbb{Z}, +)$. What is the index of $n\mathbb{Z}$ in $\mathbb{Z}$?
3. Describe in GAP the tetrahedron group $S_4$ as a subgroup of the cube group $W$ (see Fig. 3.3 on page 35). What is the index of $S_4$ in $W$?
4. It holds that $(\mathbb{Z}, +) < (\mathbb{R}, +)$. What do the left cosets of $\mathbb{Z}$ in $\mathbb{R}$ look like? How many left cosets are there?
5. Let $m \in \mathbb{N}$ and $m \geq 2$. The elements of the group $\mathbb{Z}_m^*$ are all numbers coprime to $m$ between 1 and $m - 1$ with multiplication modulo $m$ (two natural numbers are called *coprime* if their greatest common divisor is 1). The group $\mathbb{Z}_m^*$ is called the *multiplicative group of integers* mod $m$. Show with the help of this group and with Corollary 3.21 the following:
   **Euler's theorem:** *Let $a \in \mathbb{N}$ be coprime to $m \in \mathbb{N}$. Then:*

$$a^{\varphi(m)} \equiv 1 \bmod m,$$

   *where $\varphi(m)$ is Euler's totient function, i.e., the number of integers coprime to $m$ between 1 and $m - 1$.*
   (LEONHARD EULER [1707–1783] did not know groups and saw this as a purely number-theoretic theorem.)
6. Let $F$ be a regular hexagon and $U$ the stabilizer of two opposite edges of $F$ in the associated symmetry group $D_6$. To which known group is $U$ isomorphic? What is the index of $U$ in $D_6$? Determine all cosets.
7. Let $G$ be a group, $H < G$ and $g_1, g_2 \in G$. Show that $g_1 H = g_2 H$ if and only if $g_1^{-1} g_2 \in H$.
8. Let $U, V$ be finite subgroups of the group $G$ with coprime orders. Show that $U \cap V = \{e\}$.

## 3.3 Homomorphisms

In Sect. 2.3 we referred to groups as "equal", or isomorphic, when they only differ by the notation of their elements and by the appearance of the operation sign. For example, we can write the group $D_3$ as a permutation group: $D_3' = \{(), (1, 2, 3), (1, 3, 2), (1, 2), (1, 3), (2, 3)\}$. The operation is in this case the composition of permutations. On the other hand, we have $D_3 = \{id, d_{120}, d_{240}, s_a, s_b, s_c\}$ as the symmetry group of the regular triangle in the plane.

**Definition 3.22** Two groups $(G, \cdot)$ and $(H, \#)$ are called *isomorphic* if there is a bijective mapping $\phi \colon G \to H$ such that

$$\phi(u \cdot v) = \phi(u) \# \phi(v), \quad \forall u, v \in G. \tag{3.2}$$

The map $\phi$ is called an *isomorphism*, and we write $G \cong H$.

In the example of the isomorphism $\phi \colon D_3 \to D_3'$, for a given isometry that maps the triangle onto itself, the corresponding vertex permutation is taken as the image. So

$$\phi(id) = (), \ \phi(d_{120}) = (1, 2, 3), \ \phi(d_{240}) = (1, 3, 2),$$

$$\phi(s_a) = (1, 2), \ \phi(s_b) = (1, 3), \ \phi(s_c) = (2, 3).$$

Let $\mathbb{R}$ be the real line and $t$ a translation in the positive direction along this line by the distance 1. For $k \in \mathbb{Z}$, the translation $kt$ is a translation by the distance $|k|$ in the positive or negative direction, depending on whether $k$ is positive or negative. Now, the set of translations $trans = \{kt \mid k \in \mathbb{Z}\}$ forms a group $(trans, \circ)$ with respect to composition. There is an isomorphism $\phi \colon \mathbb{Z} \to trans$ from the group $(\mathbb{Z}, +)$ to $(trans, \circ)$ given by $\phi(k) = kt$. It is immediately apparent that $\phi$ is bijective and $\phi(i + j) = \phi(i) \circ \phi(j)$. Two translations add up on the line like normal numbers. The new name $trans$ is superfluous, we can simply call this group $\mathbb{Z}$.

The positive real numbers $\mathbb{R}^+$ form a group $(\mathbb{R}^+, \cdot)$ with the ordinary multiplication. The identity element is 1, the inverse of $x$ is $1/x$. There is a bijective mapping $\phi \colon \mathbb{R} \to \mathbb{R}^+$ from the group $(\mathbb{R}, +)$ to this group $(\mathbb{R}^+, \cdot)$ defined by $\phi(x) = e^x$. Because

$$\phi(x + y) = e^{x+y} = e^x e^y = \phi(x)\phi(y)$$

(here e is Euler's number $2.718\ldots$), it is an isomorphism. The inverse map $\phi^{-1}$ is the natural logarithm.

**Theorem 3.23** *Let $p$ be a prime number. Then there is (up to isomorphism) exactly one group of order $p$, the group $(\mathbb{Z}_p, +_p)$.*

**Proof** Let $G$ be any group of order $p$ and $g \in G$ with $g \neq e$. According to Corollary 3.20, $G$ is cyclic, and $g$ generates $G$. So $G = \langle g \rangle = \{id, g, g^2, \ldots, g^{p-1}\}$. The mapping $\phi \colon G \to \mathbb{Z}_p$, which is given by $g^k \to k$, is an isomorphism, as one can easily see.                                                                                               $\square$

The function `IsomorphismGroups` in GAP constructs an isomorphism, provided the two given groups are isomorphic. In the following example, we draw a square into the regular 8-gon of Fig. 3.2 on page 35 (it has vertices 1,3,5,7) and describe the stabilizer of the vertices of the square in the symmetry group of the regular 8-gon. Every isometry of the square is also an isometry of the 8-gon, and so we have the group $D_4$ as a subgroup $G$ of the group $D_8$. As generators of $G$ we have $(1,3,5,7)(2,4,6,8)$, a rotation by $90°$, and $(1,3)(8,4)(7,5)$, the reflection over the line through the vertices 2 and 6. We form the isomorphism from $G$ to the group $D_4$ by mapping these generators to a rotation and a reflection of a square with vertex labels 1,2,3,4.

```
gap> G:=Group((1,3,5,7)(2,4,6,8),(1,3)(8,4)(7,5));;
gap> H:=Group((1,2,3,4),(1,3));;
gap> f:=IsomorphismGroups(G,H);
[ (1,3,5,7), (1,3)(4,8)(5,7) ] -> [ (1,2,3,4), (1,2)(3,4) ]
```

GAP only outputs the images of the generators, as every other element can be written as a product of the generators and the condition (3.2) determines the images of all other elements: For example, given an isomorphism $g\colon G \to H$ with $G = \langle a, b \rangle$, for the element $ab^2a^{-2}$ we have

$$g(ab^2a^{-2}) = g(a)g(b)^2g(a)^{-2},$$

and solely by knowing $g(a)$ and $g(b)$ we can determine the image of $ab^2a^{-2}$. We just have to prove $g(a^{-1}) = g(a)^{-1}$, which is carried out further below.

Continuing in GAP: We can apply the isomorphism f generated in this way to an element of $G$:

```
gap> Image(f, (3,5)(2,6)(1,7));
(1,4)(2,3)
```

An isomorphism of a group onto itself is called an *automorphism*. Every group allows certain automorphisms: If $G$ is a group and $h \in G$, then the mapping $\phi_h\colon G \to G$ defined by $\phi_h(g) = h^{-1}gh$ is an automorphism, a so-called *inner automorphism*. Indeed

$$\phi_h(gg') = h^{-1}gg'h = h^{-1}gh \cdot h^{-1}g'h = \phi_h(g)\phi_h(g'),$$

which proves the condition (3.2). $\phi_h$ is also bijective, because it has the mapping $\phi_{h^{-1}}$ as its inverse. The mapping $\phi_h(g) = h^{-1}gh$ is called *conjugation* of $g$ with $h$. In an abelian group, every inner automorphism is the identity map. In the group $D_3$ with the notations of Fig. 1.1, for example,

$$\phi_{s_a}(d_{120}) = d_{240}, \quad \phi_{s_a}(s_a) = s_a, \quad \phi_{s_a}(s_b) = s_c.$$

If an automorphism is not an inner automorphism, it is called an *outer automorphism*. The automorphism $\psi\colon \mathbb{Z}_m \to \mathbb{Z}_m$ defined by $\psi(k) = m - k$ is an outer automorphism, because $\mathbb{Z}_m$ is abelian and $\psi$ is different from the identity.

The set of all automorphisms of a given group $G$ form the *automorphism group* $\mathrm{Aut}(G)$ of a group. The composition of two automorphisms is again an automorphism, and the inverse automorphism is the inverse mapping. The identity map, i.e. the automorphism which maps every group element to itself, is the identity.

GAP can calculate automorphism groups. Here is an example for the group $\mathbb{Z}_5$:

```
gap> G:=Group((1,2,3,4,5));;
gap> au:=AutomorphismGroup(G);
<group with 1 generators>
gap> Elements(au);
[ IdentityMapping( Group([ (1,2,3,4,5) ]) ),
  [ (1,2,3,4,5) ] -> [ (1,3,5,2,4) ],
  [ (1,2,3,4,5) ] -> [ (1,4,2,5,3) ],
  [ (1,2,3,4,5) ] -> [ (1,5,4,3,2) ] ]
```

The rotation by $2\pi/5$ can be mapped to any rotation, except for the rotation by $0°$, the identity. (Exercise: Prove that $\mathrm{Aut}(\mathbb{Z}_5)$ is isomorphic to $\mathbb{Z}_4$. Compare with Exercise 7.)

**Example 3.24** We have $\mathrm{Aut}(S_3) = S_3$ because: Every automorphism preserves the order of its elements. Since $(1, 2), (2, 3), (1, 3)$ are the only elements of order 2, these are permuted by an automorphism. However, any permutation of these elements determines an automorphism of the group $S_3$.

For a mapping between groups, if one only requires that the image of a product is equal to the product of the images (i.e., property (3.2)), but not bijectivity, then one obtains a homomorphism:

**Definition 3.25** Let $(G, \cdot)$ and $(H, \#)$ be groups. A mapping $\phi\colon G \to H$ is called a *homomorphism* if $\phi(u \cdot v) = \phi(u) \# \phi(v)$, $\forall u, v \in G$.

We consider the groups $(\mathbb{Z}, +)$ and $(D_7, \circ)$. We obtain a homomorphism $\phi\colon \mathbb{Z} \to D_7$ by mapping each integer $n \in \mathbb{Z}$ to the rotation by $n \cdot 360/7$ degrees in the regular 7-gon. This rotation is called $d_{n \cdot 360/7}$. The homomorphism condition is fulfilled because

$$\phi(n + m) = d_{(n+m) \cdot 360/7} = d_{n \cdot 360/7} \circ d_{m \cdot 360/7} = \phi(n) \circ \phi(m).$$

**Example 3.26** On page 29 we considered $S_4$, the symmetry group of the tetrahedron. We map the group through a homomorphism hom to the group $D_3$ by specifying the image of each generator of $S_4$.

```
gap> S4 := Group((2,4),(1,2),(1,3),(1,4),(2,3),(3,4));
Group([ (2,4), (1,2), (1,3), (1,4), (2,3), (3,4) ])
gap> D3 := Group((1,2,3),(1,2));;
gap> hom := GroupHomomorphismByImages( S4, D3,
> GeneratorsOfGroup(S4),
[(1,2),(2,3),(1,2),(1,3),(1,3),(2,3)]);
[ (2,4), (1,2), (1,3), (1,4), (2,3), (3,4) ] ->
[ (1,2), (2,3), (1,2), (1,3), (1,3), (2,3) ]
```

We can also apply the homomorphism hom      gap> Image(hom,(1,2,4));
generated in this way to pre-images.              (1,3,2)

Let $\phi: G \to H$ be any homomorphism, and $a \in G$. Denote by $e$ the identity element in $G$, and by $e'$ the identity element in $H$. It follows that $\phi(a) = \phi(e \circ a) = \phi(e)\phi(a)$. So $\phi(e) = e'$ must hold.

```
gap> Image(hom, () );
()
```

Furthermore, $e' = \phi(e) = \phi(a \circ a^{-1}) = \phi(a)\phi(a^{-1})$ and thus $\phi(a^{-1}) = \phi(a)^{-1}$. So we have:

**Theorem 3.27** *Every homomorphism $\phi: G \to H$ maps the identity element to the identity element and the inverse of an element to the inverse of its image, so for $a \in G$, $\phi(a^{-1}) = \phi(a)^{-1}$.*

The element (1,4,2) is the inverse of (1,2,4):

```
gap> Image(hom,(1,4,2)); Image(hom,(1,2,4));
(1,2,3)
(1,3,2)
```

**Example 3.28** We define a map $f: \mathcal{E} \to \{+1, -1\}$. The number $+1$ is assigned to each orientation-preserving isometry and the number $-1$ to each orientation-reversing isometry. $\{+1, -1\}$ forms a group with the usual multiplication. $f$ is a homomorphism.

To prove this statement, one distinguishes four cases. For example, the product of two orientation-reversing isometries is orientation-preserving, and this corresponds to
$$(-1) \cdot (-1) = +1.$$

**Definition 3.29** The *kernel* of a homomorphism $\phi: G \to H$ consists of all pre-images of the identity element $e' \in H$, i.e.

$$\ker(\phi) = \{g \in G \,|\, \phi(g) = e'\}.$$

The *image* of a homomorphism $\phi: G \to H$ is the set of all elements that appear as an image, i.e.

$$\text{im}(\phi) = \{h \in H \,|\, \exists g \in G, \phi(g) = h\}.$$

The kernel of the homomorphism $f$ of Example 3.28 is the subgroup of orientation-preserving isometries of $\mathcal{E}$.

**Theorem 3.30** *Let $\phi: G \to H$ be a homomorphism. Then $\ker(\phi)$ forms a subgroup of $G$ and $\text{im}(\phi)$ forms a subgroup of $H$.*

**Proof** According to Theorem 3.9 on page 35 we have to check for all $a, b \in \ker(\phi)$ that $ab^{-1} \in \ker(\phi)$. The statement $a, b \in \ker(\phi)$ means that $\phi(a) = e'$ and $\phi(b) = e'$, where $e'$ is the identity element of $H$. From Theorem 3.27 it follows that $\phi(b^{-1}) = \phi(b)^{-1} = e'^{-1} = e'$ and therefore $e' = \phi(a)\phi(b^{-1}) = \phi(ab^{-1})$, i.e. $ab^{-1} \in \ker(\phi)$.

If $g', h' \in \text{im}(\phi)$, then there exist $g, h \in G$ with $\phi(g) = g'$, $\phi(h) = h'$. Then $\phi(h^{-1}) = h'^{-1}$ and $\phi(gh^{-1}) = g'h'^{-1} \in \text{im}(\phi)$.                                                                  □

We consider the kernel of the homomorphism of Example 3.26:

hom: $S_4 \rightarrow D_3$

```
gap> Kernel( hom );
Group([ (1,4)(2,3), (1,3)(2,4) ])
gap> Elements(last);
[ (), (1,2)(3,4), (1,3)(2,4),
  (1,4)(2,3) ]
```

The kernel is isomorphic to the Klein four-group, the symmetry group of a rectangle, where the successive vertices are numbered 1,3,2,4. The entire image of the homomorphism is obtained with the following command:

```
gap> Image(hom);
Group([ (1,2), (2,3), (1,2), (1,3), (1,3), (2,3) ])
gap> Elements(last);
[ (), (2,3), (1,2), (1,2,3), (1,3,2), (1,3) ]
```

The homomorphism hom is surjective. The image is the entire group $D_3$.

The identity element is always contained in the kernel of a homomorphism, as we proved in Theorem 3.27. If the kernel consists only of the identity element, then we call the kernel of the homomorphism *trivial*.

**Theorem 3.31** *Let $\phi: G \rightarrow H$ be a homomorphism. Then $\phi$ has a trivial kernel if and only if $\phi$ is injective.*

**Proof** If $\phi$ is not injective, then there exist $g_1, g_2 \in G$ with $g_1 \neq g_2$, but $\phi(g_1) = \phi(g_2)$. Then $\phi(g_1g_2^{-1}) = \phi(g_1)\phi(g_2)^{-1} = 1$, i.e., $g_1g_2^{-1}$ is in $\ker(\phi)$. But $g_1g_2^{-1} \neq 1$, because $g_1 \neq g_2$.

Conversely, if $\phi$ is injective, then the identity element of $H$ can only have one preimage. This means that the kernel of $\phi$ is trivial.                                      □

If $G < H$, then the map $\phi: G \rightarrow H$ which maps each element of $G$ to its corresponding image in $H$ is an injective homomorphism. This homomorphism is called the *embedding* of $G$ into $H$.

**Theorem 3.32** *Let $\phi: G \rightarrow H$ be a homomorphism and $U < H$. Then $\phi^{-1}(U)$ is a subgroup of $G$.*

**Proof** If $g, g' \in \phi^{-1}(U)$, then $\phi(g), \phi(g') \in U$. Then we also have $\phi(g)\phi(g')^{-1} \in U$, and because $\phi(gg'^{-1}) = \phi(g)\phi(g')^{-1}$, it follows that $gg'^{-1} \in \phi^{-1}(U)$.

According to the subgroup criterion (Theorem 3.9), it follows that $\phi^{-1}(U) < G$.                                      □

## Exercises

1. Let $U$ be a subgroup of the group $G$ and $g \in G$. Prove that the conjugate subgroup $gUg^{-1}$ has the same order as $U$.
2. Let $\phi: G \rightarrow H$ be an isomorphism and $x \in G$. Show that $x$ and $\phi(x)$ have the same order.
3. Prove that $O_2$ is isomorphic to the symmetry group of a circle in the plane.

4. We define $n\mathbb{Z} = \{k \cdot n \mid k \in \mathbb{Z}\}$ for any $n \in \mathbb{N}$. Show that the group $(\mathbb{Z}, +)$ is isomorphic to the group $(n\mathbb{Z}, +)$ for any $n \in \mathbb{N}$.
5. Prove that $\forall n \in \mathbb{Z}$ the mapping $\phi_n : \mathbb{Z} \to \mathbb{Z}$ defined by $\phi_n(m) = n \cdot m$ is a group homomorphism. The mapping $\psi_n : \mathbb{Z} \to \mathbb{Z}$ defined by $\psi_n(m) = n + m$ is not a group homomorphism.
6. Determine the automorphism group of $\mathbb{Z}$.
7. Determine the automorphism group of $\mathbb{Z}_n$. Note that an automorphism of $\mathbb{Z}_n$ must map elements to elements of the same order. A generator of $\mathbb{Z}_n$ must therefore be mapped to an element of order $n$. So first determine the elements of order $n$ of $\mathbb{Z}_n$ (possibly with the help of GAP).
8. Let $n, m \in \mathbb{N}$ be such that $m$ is a divisor of $n$. Show that there exists a subgroup $U < \mathbb{Z}_n$ with $U \cong \mathbb{Z}_m$.

## 3.4  Normal Subgroups

We consider the group $(\mathbb{Z}, +)$ and the subgroup

$$7\mathbb{Z} = \{\ldots, -14, -7, 0, 7, 14, 21, \ldots\} = \{7 \cdot k \mid k \in \mathbb{Z}\}$$

(see also Exercise 2. of Sect. 3.2). According to Theorem 3.12, $(\mathbb{Z}, +)$ can be disjointly decomposed into a union of (left) cosets. The cosets here are:

$$7\mathbb{Z}, \ 1 + 7\mathbb{Z}, \ 2 + 7\mathbb{Z}, \ 3 + 7\mathbb{Z}, \ 4 + 7\mathbb{Z}, \ 5 + 7\mathbb{Z}, \ 6 + 7\mathbb{Z}. \tag{3.3}$$

Here, the cosets can even be combined with normal addition: For example, adding $15 = 1 + 2 \cdot 7 \in 1 + 7\mathbb{Z}$ and $26 = 5 + 3 \cdot 7 \in 5 + 7\mathbb{Z}$ gives $15 + 26 = 41 = 6 + 5 \cdot 7 \in 6 + 7\mathbb{Z}$. We have $(1 + 7\mathbb{Z}) + (5 + 7\mathbb{Z}) = 6 + 7\mathbb{Z}$, or more generally

$$(k + 7\mathbb{Z}) + (j + 7\mathbb{Z}) = (k + j) + 7\mathbb{Z}$$

(proving this relationship is not difficult). The 7 cosets from (3.3) can therefore be considered as group elements of a group that "inherits" its group operation from the group $(\mathbb{Z}, +)$.

In general, let $N$ be a subgroup of the group $(G, \circ)$ with cosets $g_1 N$, $g_2 N$, $g_3 N, \ldots$. The question is: Can the cosets be combined using the operation of $G$? Can one define an operation "$\cdot$" of left cosets as

$$g_i N \cdot g_k N = (g_i \cdot g_k) N,$$

so that a group of cosets is formed (each group element should be exactly one coset)? This is possible if the operation is independent of the representatives of the cosets, i.e. if

$$h_i \in g_i N \text{ and } h_k \in g_k N \Rightarrow (g_i \cdot g_k)N = (h_i \cdot h_k)N. \qquad (3.4)$$

We will soon find out that this is not always possible, but let's assume it for now. It then follows that

$$h_i \in g_i N \text{ and } h_k \in g_k N \Rightarrow h_i \cdot h_k \in (g_i \cdot g_k)N$$

or, in other words, $\forall g_i, g_k \in G, \ \forall n_i, n_k \in N$ we have

$$h_i = g_i \cdot n_i \text{ and } h_k = g_k \cdot n_k \Rightarrow \exists w \in N \text{ with } h_i \cdot h_k = g_i \cdot g_k \cdot w,$$

i.e.

$$(g_i \cdot n_i) \cdot (g_k \cdot n_k) = g_i \cdot g_k \cdot w.$$

We cancel $g_i$ from the left, giving

$$n_i \cdot g_k \cdot n_k = g_k \cdot w.$$

Thus $n_i \cdot g_k = g_k \cdot w \cdot n_k^{-1}$, with $w \cdot n_k^{-1} \in N$. For all $n_i \in N$ it follows that $n_i \cdot g_k \in g_k N$, i.e.

$$Ng_k \subset g_k N. \qquad (3.5)$$

If you replace $g_k$ with $g_k^{-1}$, you get $n_i \cdot g_k^{-1} = g_k^{-1} \cdot w \cdot n_k^{-1}$ or $g_k \cdot n_i = w \cdot n_k^{-1} \cdot g_k$ and thus

$$g_k N \subset Ng_k. \qquad (3.6)$$

From (3.5) and (3.6) it follows that

$$g_k N = Ng_k. \qquad (3.7)$$

So left cosets are equal to right cosets.

**Definition 3.33** Let $N$ be a subgroup of a group $G$. If for all $g \in G$ the relation $gN = Ng$ holds, then $N$ is called a *normal subgroup* of $G$. We use the notation $N \lhd G$.

It is easy to see that in an abelian group every subgroup is normal.

In Example 3.11 we considered the subgroup $D_4^+$ of the group $D_4$. Here left and right cosets are equal (although we have not yet checked this for all $g \in D_4$), so that $D_4^+ \lhd D_4$ (compare Exercise 3.). 

In GAP you get all normal subgroups of a group with the command NormalSubgroups. In the following we generate the group $D_4$ by a rotation and a reflection.

```
gap> D4:=Group((1,3),(1,2,3,4));;
gap> Ns:=NormalSubgroups(D4);
[ Group(()), Group([ (1,3)(2,4) ]),
  Group([ (1,3)(2,4), (1,4)(2,3) ]),
  Group([ (1,3)(2,4), (1,2,3,4) ]), Group([ (2,4),
  (1,3) ]),
  Group([ (1,3), (1,2,3,4) ]) ]
gap> List(Ns,Order);
[ 1, 2, 4, 4, 4, 8 ]
```

With List(Ns,Order); we get the orders of all normal subgroups. The trivial group and the whole $D_4$ are normal subgroups of orders 1 and 8. The normal subgroup of order 2 consists of the rotation by 180° and the identity. The last normal subgroup of order 4 is generated by two perpendicular mirror axes and is thus isomorphic to the group of the rhombus.

The normal subgroup Group([ (1,3)(2,4), (1,2,3,4) ]) consists of all rotations and is thus $D_4^+$. However, this subgroup can also be generated by just one element:

```
gap> MinimalGeneratingSet(Ns[4]);
[ (1,2,3,4) ]
```

A bit more theory: For a subgroup $U$ of the group $G$, we denote the set of cosets of $U$ by $G/U$. Each individual element of $G/U$ is thus a coset. Sometimes we speak of $G$ *modulo* $U$.

**Theorem 3.34** *Let $N < G$. Then $N$ is a normal subgroup of $G$ if and only if the cosets $gN$ for all $g \in G$ form a group $G/N$ with the operation*

$$g_i N \cdot g_k N = (g_i g_k) N.$$

***Proof*** We have already proven the equivalence in one direction. It remains to show that if $N \lhd G$ then the cosets form a group. So we have to show that condition (3.4) holds. Let $h_i \in g_i N$ and $h_k \in g_k N$. We have

$$h_i h_k N = h_i g_k N \text{ because } h_k \in g_k N$$

$$= h_i N g_k \text{ because } g_k N = N g_k$$

$$= g_i N g_k \text{ because } h_i \in g_i N$$

$$= g_i g_k N \text{ because } g_k N = N g_k.$$

The product $g_i N \cdot g_k N = (g_i g_k) N$ is therefore well-defined, and so $G/N$ forms a group.                                                                                           □

**Definition 3.35** Let $N \lhd G$. The group $G/N$, in which the elements are the cosets $gN$ and the operation is determined by $g_i N \cdot g_k N = (g_i g_k) N$, is called the *factor group* of $G$ mod $N$. It is also called the *quotient* of the group $G$ by $N$.

The identity element of the factor group $G/N$ is $N$, because $N \cdot gN = gN \cdot N = gN$. The inverse of $gN$ is $g^{-1}N$, because $gN \cdot g^{-1}N = gg^{-1}N = N$.

Let $N$ be a subgroup of an abelian group $G$. Then $N$ is a normal subgroup. The factor group $G/N$ is then also abelian, because $g_i N \cdot g_k N = (g_i g_k) N = (g_k g_i) N = g_k N \cdot g_i N$.

Back to our example $(\mathbb{Z}, +)$ with the subgroup $7\mathbb{Z}$: The group $\mathbb{Z}/7\mathbb{Z}$ has the elements

$$\mathbb{Z}/7\mathbb{Z} = \{7\mathbb{Z}, \ 1 + 7\mathbb{Z}, \ 2 + 7\mathbb{Z}, \ 3 + 7\mathbb{Z}, \ 4 + 7\mathbb{Z}, \ 5 + 7\mathbb{Z}, \ 6 + 7\mathbb{Z}\}.$$

It is isomorphic to the group $(\mathbb{Z}_7, +_7)$, because the addition in $\mathbb{Z}/7\mathbb{Z}$ behaves like addition mod 7:

$$(3 + 7\mathbb{Z}) + (5 + 7\mathbb{Z}) = 1 + 7\mathbb{Z}.$$

In $(\mathbb{Z}_7, +_7)$ we calculate $3 +_7 5 = 1$.

**Definition 3.36** A group $G$ is called *residually finite* if for every $g \in G$, $g \neq 1$, there is a finite quotient in which $g$ is nontrivial.

Equivalent to this is the following definition: $G$ is called *residually finite* if the intersection of all normal subgroups of finite index consists only of the trivial element.

Of course, every finite group is residually finite. For a finite group $G$ take $G/\{e\}$ as the finite quotient where $e$ is the identity element of $G$.

**Theorem 3.37** $\mathbb{Z}$ *is residually finite.*

**Proof** For a number $n \in \mathbb{Z}$, $n \neq 0$, choose a prime number $p \in \mathbb{N}$ which is coprime to $|n|$. Then $n$ is nontrivial in the quotient $\mathbb{Z}/p\mathbb{Z} = \mathbb{Z}_p$, because $n$ is in the corresponding residue class $n + p\mathbb{Z}$, and this class is not $p\mathbb{Z}$, because $n$ is coprime to $p$.                                                                                                       □

In Appendix B, residual finiteness is proven for another class of groups, namely the groups $\mathrm{GL}(n, \mathbb{Z})$ of all invertible $n \times n$ matrices with entries in $\mathbb{Z}$.

The normal subgroup property $gN = Ng$ is an equation between two sets. Obviously equivalent to this is $gNg^{-1} = N$. It is even sufficient to demand $gNg^{-1} \subset N$ for all $g \in G$, because if $gng^{-1} = n' \in N$, then $n = g^{-1}n'g \in g^{-1}N(g^{-1})^{-1}$ for all $n \in N$. For $h = g^{-1}$ we have $N \subset hNh^{-1}$. We have proven:

**Lemma 3.38** *Let $N$ be a subgroup of the group $G$. If $gNg^{-1} \subset N$ for all $g \in G$, then $N$ is a normal subgroup of $G$.*

**Theorem 3.39** *The kernel of a homomorphism* $\phi \colon G \to H$ *is a normal subgroup of the group* $G$.

***Proof*** In Theorem 3.30 we have shown that $\ker(\phi)$ is a subgroup of $G$. Only the normal subgroup property is missing.

According to Lemma 3.38 we have to show $g \ker(\phi) g^{-1} \subset \ker(\phi)$ for all $g \in G$. Let $a \in \ker(\phi)$, i.e. $\phi(a) = 1$. Let $g \in G$. Then

$$\phi(gag^{-1}) = \phi(g)\phi(a)\phi(g^{-1}) = \phi(g)1\phi(g)^{-1} = 1,$$

i.e. $gag^{-1} \in \ker(\phi)$. □

The converse also holds: Every normal subgroup can be considered as the kernel of a homomorphism:

**Theorem 3.40** *Let* $N \lhd G$. *Then* $N = \ker(\phi)$ *for a homomorphism* $\phi \colon G \to H$, *the so-called* canonical homomorphism. *Here,* $H = G/N$.

***Proof*** The homomorphism $\phi \colon G \to G/N$ maps $N$ to the identity element. An element $g \notin N$ is mapped to the coset $gN \neq N$. □

This theorem can be formulated differently:

**Theorem 3.41 (1st Isomorphism Theorem)** *If* $\phi \colon G \to H$ *is a homomorphism, then* $\mathrm{im}(\phi)$ *is isomorphic to* $G/\ker(\phi)$.

***Proof*** Let $N = \ker(\phi)$ and $\psi \colon G/N \to \mathrm{im}(\phi)$ be the mapping that maps $gN \to \phi(g)$. To ensure that this mapping is well-defined and bijective, we have to show $\phi(a) = \phi(b) \iff aN = bN$. We assume $\phi(a) = \phi(b)$. Then $1 = (\phi(b))^{-1}\phi(a) = \phi(b^{-1})\phi(a) = \phi(b^{-1}a)$. Thus, $b^{-1}a \in N$ and therefore $N = b^{-1}aN$. But this is the same as $aN = bN$. The converse is obtained by running the argument backwards.

We still have to show that $\psi$ is an isomorphism, i.e., satisfies condition (3.2). We have

$$\psi(aNbN) = \psi(abN) = \phi(ab) = \phi(a)\phi(b) = \psi(aN)\psi(bN).$$

□

We calculate an example in GAP.

We consider the symmetric group $S_4$ and a subgroup that is isomorphic to the Klein four-group.

```
gap> S4:=SymmetricGroup(4);;
gap> V:=Subgroup(S4,[(1,2)(3,4),(1,3)(2,4)]);
Group([ (1,2)(3,4), (1,3)(2,4) ])
gap> Elements(V);
[ (), (1,2)(3,4), (1,3)(2,4), (1,4)(2,3) ]
```

$V$ is a normal subgroup in the group $S_4$, and we can form the factor group. This has $24/4 = 6$ elements, because there are 6 cosets.

```
gap> IsNormal(S4,V);
true
gap> F:=FactorGroup(S4,V);
Group([ f1, f2 ])
gap> Order(F);
6
```

We form the associated homomorphism $hom\colon S_4 \to F = S_4/V$.

```
gap> hom:=NaturalHomomorphismByNormalSubgroup(S4,V);
[ (1,2,3,4), (1,2) ] -> [ f1*f2, f1 ]
```

$F$ is isomorphic to the group $S_3$:

```
gap> StructureDescription(F);
"S3"
```

$S_3$ is isomorphic to the group $D_3$. The elements of $F$ consist of the identity, a reflection $f1$, a rotation by $120°$ $f2$, the rotation by $240°$ $f2^2$ and reflections $f1 \circ f2$ and $f1 \circ f2^2$.

```
gap> Elements(F);
[ <identity> of ..., f1, f2, f1*f2, f2^2, f1*f2^2 ]
```

We map a few elements. $(1, 4)$ lies in the same coset as $(2, 3)$, so it is mapped to the same element. Elements of $V$ are mapped to the identity element. The kernel of *hom* is precisely the group $V$.

```
gap>  Image(hom,(1,4));
f1*f2^2
gap>  Image(hom,(2,3));
f1*f2^2
gap>  Image(hom,(1,2,3));
f2
gap> Image(hom,(1,4)(2,3));
<identity> of ...
gap> Kernel(hom);
Group([ (1,2)(3,4), (1,3)(2,4) ])
```

**Example 3.42** An example where the subgroup is not a normal subgroup: Let $G$ be the symmetry group of the regular hexagon $F$ of Fig. 3.4 and $S$ the stabilizer of the vertices 2 and 5. We play in GAP.

**Fig. 3.4** A regular hexagon in the plane

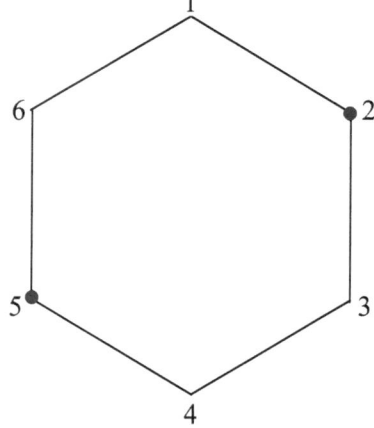

```
// The group G is generated by a reflection
// and a rotation by 60 degrees
gap> G:=Group((2,6)(3,5),(1,2,3,4,5,6));;
// S is the stabilizer of the vertices 2, 5
// generated by 2 reflections
gap> S:=Subgroup(G,[(1,3)(6,4),(1,6)(2,5)(3,4)]);
Group([ (1,3)(4,6), (1,6)(2,5)(3,4) ])
gap> Elements(S);
[ (), (1,3)(4,6), (1,4)(2,5)(3,6), (1,6)(2,5)(3,4) ]

// There are 3 cosets
gap> Index(G,S);
3

// S is not a normal subgroup
gap> IsNormal(G,S);
false

// We check this by comparing a*S with S*a
// The right and left cosets are different
gap> a:=(6,2)(5,3);;
gap> a*Elements(S);
[ (2,6)(3,5), (1,3,5)(2,4,6), (1,4)(2,3)(5,6),
  (1,6,5,4,3,2) ]
gap> Elements(S)*a;
[ (2,6)(3,5), (1,5,3)(2,6,4), (1,4)(2,3)(5,6),
  (1,2,3,4,5,6) ]
```

**Theorem 3.43** *Let G be the symmetry group of a figure in the plane (or in $\mathbb{R}^n$). Assume G contains an orientation-reversing isometry s. Then G contains exactly as many orientation-preserving as orientation-reversing isometries.*

**Proof** Let $\phi: G \to \mathbb{Z}_2$ be the homomorphism that maps orientation-preserving isometries to the identity element 0 and orientation-reversing isometries to 1. The kernel of $\phi$ is the group of orientation-preserving isometries $G^+$. The only other left coset $s \circ G^+$ is the set of orientation-reversing isometries. Because cosets are all the same size, the theorem is proven. □

**Exercises**

1. Provide a homomorphism from the group $(\mathbb{R}, +)$ to the symmetry group of the circle. Describe the elements of the kernel and the associated factor group.
2. Let $U < D_3$ be the subgroup generated by a reflection. What are the elements of $U$? Is $U$ normal in $D_3$? Determine all cosets. (Hint: Consider Fig. 1.1 on page 2, write the group $D_3$ as

$$D_3 = \{(), (1, 2, 3), (1, 3, 2), (1, 2), (1, 3), (2, 3)\}$$

with $U = \langle (1, 2) \rangle$, and check if $gUg^{-1} \subset U$ for all $g \in D_3$. If you have trouble doing it by hand, use GAP.)

3. Prove that if $H < G$ is of index 2, then $H$ is normal in $G$. Using this, show that the orientation-preserving isometries are normal in the group of all isometries.
4. Let $H < G$ and $H' \lhd G$. Show that $H \cap H'$ is normal in $G$.
5. Prove, perhaps with the help of GAP, that the subgroup $D_2$ in $D_8$ of Example 3.8 is not normal.
6. Let $U < D_6$ be the subgroup generated by the point reflection $p$ through the center of the hexagon. Then $U = \{id, p\}$. Prove that $U$ is normal in $D_6$, and show that $D_6/U \cong D_3$.
7. In the group $D_6$ let $d$ be the anti-clockwise $60°$ rotation and $U = \langle d^2 \rangle$.

   (a) Determine $|U|$ and the index of $U$ in $D_6$.
   (b) Determine the left and right cosets and show that $U$ is normal in $D_6$.
   (c) Make some calculations in the factor group.
   (d) Provide the homomorphism of $D_6$ into the factor group.

8. Show that the set of inner automorphisms of a group $G$ is a normal subgroup of $\mathrm{Aut}(G)$.

## 3.5  Translations

In Theorem 3.6 we proved that the translations $\mathcal{T}$ form a subgroup of the group of isometries of the plane. In fact, we even have:

**Theorem 3.44** *The translations form a normal subgroup in the isometry group $\mathcal{E}$ of the plane.*

**Proof** According to Lemma 3.38 we have to prove $g\mathcal{T}g^{-1} \subset \mathcal{T}$ for any isometry $g \in \mathcal{E}$. By Theorem 2.24, $\mathcal{E}$ is generated by reflections. If we can prove the relation $s\mathcal{T}s^{-1} \subset \mathcal{T}$ for all reflections $s \in \mathcal{E}$, then it will hold for all isometries $g \in \mathcal{E}$, because if $g \in \mathcal{E}$ is any isometry, then we can write it as a product of reflections $g = s_1 s_2 \ldots s_k$ and

$$
\begin{aligned}
g\mathcal{T}g^{-1} &= s_1 s_2 \ldots s_k \mathcal{T}(s_1 s_2 \ldots s_k)^{-1} \\
&= s_1 s_2 \ldots s_k \mathcal{T}s_k^{-1} \ldots s_2^{-1} s_1^{-1} \\
&\subset s_1 s_2 \ldots s_{k-1} \mathcal{T}s_{k-1}^{-1} \ldots s_2^{-1} s_1^{-1} \\
&\subset \mathcal{T}.
\end{aligned}
$$

So we show $s\mathcal{T}s^{-1} \subset \mathcal{T}$ for any reflection $s \in \mathcal{E}$: For this, we consider Fig. 3.5. Here we recognize that a reflection followed by a translation is the same as another translation followed by the same reflection. In other words, $s_a t_v = t_w s_a$, where $s_a$ is the reflection over any line $a$, $t_v$ is the translation along any vector $\vec{v}$ and $t_w$ is the translation along the vector $\vec{v}$ reflected over $a$. It follows that $s_a \mathcal{T} s_a^{-1} \subset \mathcal{T}$.  □

**Fig. 3.5** Reflection and
Translation

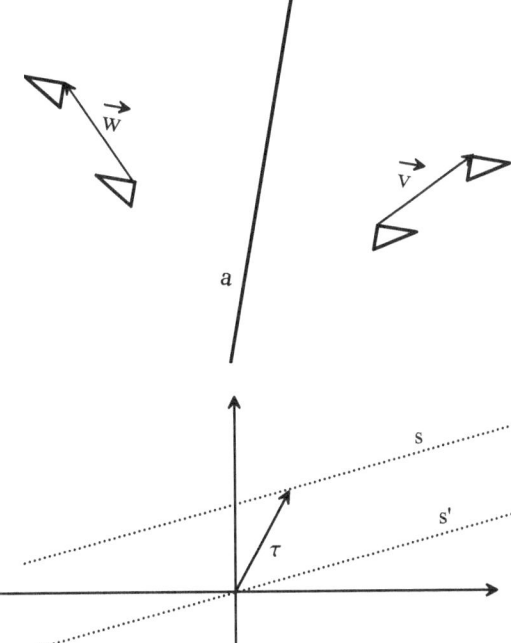

**Fig. 3.6**   $s = \tau s' \tau^{-1}$

Because of Theorem 3.40, $\mathcal{T}$ is therefore the kernel of a homomorphism $\phi$. To investigate the associated factor group, we write any element of $\mathcal{E}$ as a product of rotations around the origin, reflections over lines through the origin, and translations. We check if this is always possible. To do this, we consider any reflection $s$. The translation $\tau$ is chosen so that its inverse moves $s$ to a line $s'$ passing through the origin. Then we can replace $s$ with $\tau s' \tau^{-1}$, as can be easily seen in Fig. 3.6.

Similarly, it is clear that for any rotation $d$ there is a translation $\tau$ and a rotation $d'$ around the origin such that $d = \tau d' \tau^{-1}$.

So if we have written any element of $\mathcal{E}$ as a product of rotations around the origin, reflections over lines through the origin, and translations, we obtain the image of $\phi$ by suppressing the translations. Thus, the origin remains fixed in the image, and we have a surjective homomorphism $\phi \colon \mathcal{E} \to O_2$. The kernel consists exactly of all translations.

We have thus proven the following theorem:

**Theorem 3.45** *Let $\tau$ be the translation that maps the coordinate origin to the point $P$, and $O_P < \mathcal{E}$ be the subgroup that leaves $P$ fixed. Then $O_P = \tau O_2 \tau^{-1}$.*

We will revisit conjugation in Sect. 4.3.

Every element $g \in \mathcal{E}$ can be represented as a product of an element from $O_2$ followed by a translation: Let $P$ be the point of the plane to which the element $g$ maps the origin, i.e., $g(0) = P$. There is exactly one translation $\tau$ with $\tau(0) = P$.

Then $h = \tau^{-1} \circ g \in O_2$, and therefore $g$ can be written as $g = \tau \circ h$. This gives us the following:

**Theorem 3.46** $\mathcal{E} = \mathcal{T}O_2 = \{\tau \circ h \mid \tau \in \mathcal{T}, h \in O_2\}$ and $\mathcal{E}^+ = \mathcal{T}O_2^+$.

The second statement follows from the fact that translations are orientation-preserving.

**Theorem 3.47** $[\mathcal{E} : \mathcal{E}^+] = 2$.

**Proof** Let $s \in \mathcal{E}$ be orientation-reversing, i.e., a reflection or a glide reflection. If $s' \in \mathcal{E}$ is any orientation-reversing element, then $ss'$ is orientation-preserving, so $ss' \in \mathcal{E}^+$.

Therefore, $s' \in s\mathcal{E}^+$, and we have proven

$$\mathcal{E} = \mathcal{E}^+ \cup s\mathcal{E}^+.$$

$\square$

Together with Exercise 3. of Sect. 3.4, we now get:

**Corollary 3.48** $1 \lhd \mathcal{T} \lhd \mathcal{E}^+ \lhd \mathcal{E}$.

From Theorem 3.44, it follows that the subgroup of translations of a symmetry group $G$ of a figure $F$ is a normal subgroup in $G$.

**Example 3.49** In the symmetry group $G$ of the frieze in Fig. 1.3, let $T$ be the normal subgroup of translations. Then $[G : T] = 2$ and $G$ is isomorphic to $\mathbb{Z}$.

What do the left cosets look like here? $G$ is generated by the glide reflection $s = s_a t_v$ and the translation $\tau = t_v^2$. So $G = \langle s, \tau \rangle$, and it is commutative because the generators commute with each other, i.e. $s\tau = \tau s$, as can be easily seen in the figure. So if we want to write an element $g \in G$ in terms of the generators, we can first perform all glide reflections and then all translations. So we can write $g$ as $g = \tau^m s^k$. Two glide reflections can be replaced by a translation. So if $k$ is even, then $g = \tau^{m+k/2}$, and $g = s\tau^{m+(k-1)/2}$ otherwise. The cosets are therefore $T$ and $sT$ and $[G : T] = 2$.

Since $s^2 = \tau$ we have $G = \langle s \rangle$ and $G \cong \mathbb{Z}$.

**Exercises**

1. (a) Prove that the elements of a group $G$ that commute with all group elements form a subgroup, the *center* of $G$:

$$C(G) = \{h \in G \mid \forall g \in G \quad gh = hg\}.$$

This subgroup is even a normal subgroup. For abelian groups $G$, $C(G) = G$.

   (b) The group $D_4$ has nontrivial center ( GAP gives you $C(D_4)$ with the command gap> Centre(DihedralGroup(8)); . Prove the output of GAP). The center of the group $D_3$ is trivial.

(c)  Which dihedral groups have nontrivial center?

(d)  Show that the group $S_3$ has trivial center.

2.  Generalize the definition of the homomorphism $\phi \colon \mathcal{E} \to O_2$ to the symmetry group of $\mathbb{R}^n$.

3.  Given a rotation $d$ in the plane, find a translation $\tau$ and a rotation $d' \in O_2$ such that $d = \tau d' \tau^{-1}$.

# Chapter 4
# Group Operations

The first section deals with a special class of finite groups which have significance far beyond group theory: groups of permutations. In the two following sections, what we have been doing so far is formalized: We have so far understood a group as a set of isometries of an object. This is clarified and generalized. Groups "operate" or "act" on sets. For example, the group of the plane operates on (the set of points of) the plane. The group of the square operates on the vertices of a square. Each group element maps a vertex of the square to another. Section 4.3 deals extensively with the conjugation of elements in a group, the conjugation of subgroups, and the geometric interpretation of conjugation, as already implicitly seen in Theorem 3.45.

The last two sections bring another reinterpretation: Groups themselves become geometric objects. Then one can work with groups using the methods of geometry. This idea will be revisited in Chap. 10.

## 4.1 The Symmetric Group

The first groups that were systematically studied by Lagrange, Cauchy, and others in the first half of the nineteenth century were permutation groups.

We will see that every finite group is isomorphic to a permutation group, just like the group $D_3$ was already written as a permutation group

$$D_3 = \{(), (1, 2, 3), (1, 3, 2), (1, 2), (1, 3), (2, 3)\}. \tag{4.1}$$

We formulate more precisely what we mean by a permutation group. Let $T_n = \{1, 2, \ldots, n\}$ be the set of natural numbers from 1 to $n$ for any $n > 1$. A bijective mapping of $T_n$ onto itself is called a *permutation*. Let $S_n$ be the set

© The Author(s), under exclusive license to Springer-Verlag GmbH, DE, part of Springer Nature 2024
S. Rosebrock, *Visual Group Theory*, Springer Undergraduate Mathematics Series,
https://doi.org/10.1007/978-3-662-69365-0_4

of all permutations of $T_n$. If you compose 2 permutations, then you again get a permutation. For example: $(2, 4, 5)(1, 3) \circ (1, 2, 3) = (1, 4, 5, 2)$ or

$$(1, 2, 3, \ldots, n)^n = (1)(2) \ldots (n) = id. \tag{4.2}$$

In GAP:
```
gap> (1,2,3,4,5)^5;
()
```

A set of permutations that form a group with respect to composition is called a *permutation group*. It doesn't matter whether the elements of the set $T_n$ are permuted or are replaced by any other $n$ different elements. The resulting group is ultimately the "same".

Permutations are represented by compositions of cycles. A *cycle of length m* (for short, an *m-cycle*) is a bracket expression of the form $(a_1, a_2, \ldots, a_m)$, where the $a_i \in T_n$ are all distinct. The expression means that $a_i$ is mapped to $a_{i+1}$ (in the index mod $m$).

The elements of $S_n$ can be written as products of disjoint cycles, such as $(1, 5, 3)(2, 6)$. This decomposition uniquely indicates the number to which each element of $T_n$ is mapped. Cycles of length 2 are called *transpositions*.

**Theorem 4.1** $S_n$ *forms a group for* $n > 1$ *with respect to composition.*

**Proof** The associativity for functions is always proven in the same way, namely as in the proof of Example 2.2 on page 15. The composition of functions is always associative. The identity element is the identical permutation. Every element of $T_n$ is mapped to itself. In the above notation: $id = (1)(2) \ldots (n)$ or $id = ()$. The inverse of a bijective mapping is again bijective, thus also a permutation. □

$S_n$ is called the *symmetric group over n elements*. More generally: If you have any (possibly infinite) set $X$, $S_X$ denotes the group of permutations of $X$ with respect to composition. What is the order of $S_n$? The element 1 can be mapped to $n$ different numbers, 2 can then only be mapped to $n - 1$ numbers etc., i.e. $|S_n| = n! = n \cdot (n - 1) \cdot \ldots \cdot 2 \cdot 1$.

The order of an $m$-cycle in $S_n$ is $m$, as we can see directly from Formula (4.2).

The groups $S_n$ are generally not abelian, for example

$$(1, 3) \circ (1, 2) = (1, 2, 3) \neq (1, 3, 2) = (1, 2) \circ (1, 3)$$

in $S_3$. We have $S_{n-1} < S_n$, because all permutations of $S_n$ that fix the element $n$ (i.e. map it to itself), form a subgroup that exactly permutes all numbers from 1 to $n - 1$, so it is isomorphic to $S_{n-1}$. So $S_2 < S_3 < S_4 < \cdots$, and since the group $S_3$ is not abelian, all $S_n$ for $n \geq 3$ are not abelian.

In Sect. 2.5 we have already determined that the order of a permutation is equal to the least common multiple of the cycle lengths. So every $m$-cycle generates a subgroup isomorphic to $\mathbb{Z}_m$.

We have $S_3 = D_3$, because the group $D_3$ is represented in (4.1) as all permutations of 3 elements.

We have extensively considered the tetrahedron group $S_4$ in Example 2.22 on page 27. A renewed look at that example is helpful at this point. Every permutation of the vertices of a tetrahedron leads to an isometry of the entire tetrahedron. Therefore, the symmetry group of the tetrahedron is isomorphic to the group $S_4$ of permutations. In Example 2.22 we found that the group $S_4$ is generated by transpositions. This applies more generally:

**Theorem 4.2** *The group $S_n$ for $n > 1$ is generated by transpositions.*

**Proof** We have

$$(1, m) \circ (1, m - 1) \circ \cdots \circ (1, 4) \circ (1, 3) \circ (1, 2) = (1, 2, 3, 4, \ldots, m).$$

Therefore, we can represent every $m$-cycle for $m \geq 3$ as a product of transpositions. Every $m$-cycle is a product of $m - 1$ transpositions. Since every permutation can be represented as a product of cycles, we can therefore write every permutation as a product of transpositions. □

For example:
```
gap> (4,7)*(4,2)*(4,6)*(4,1)*(4,3);
(1,3,4,7,2,6)
```
This is equal to the permutation (4,7,2,6,1,3).

The following theorem of CAYLEY states that *every* finite group can be written as a group of permutations:

**Theorem 4.3** *Every group of order $n$ is isomorphic to a subgroup of the group $S_n$.*

**Proof** Let $G = \{a_0, a_1, \ldots, a_{n-1}\}$ be the given group with $n$ elements. In the proof of Theorem 3.14 on page 38 we saw that left multiplication by an element $a_i$, i.e. the mapping $\phi_i : G \to G$ defined by $\phi_i(g) = a_i g$, is a bijective mapping. (Caution: This map is in general not an isomorphism.) So we have a bijective map from $G$ to itself, in other words, a permutation of the elements of $G$.

The set $\Phi = \{\phi_0, \phi_1, \ldots, \phi_{n-1}\}$ of permutations of $n$ elements (the elements of $G$) forms a subgroup of $S_n$, because $\forall g \in G$ we have

$$\phi_i \phi_k(g) = a_i a_k g = a_t g = \phi_t(g), \text{ if } a_i a_k = a_t.$$

The composition of these permutations is thus closed in the set $\Phi$. If $a_0$ is the identity element of $G$, then $\phi_0$ is the identity element of $\Phi$, because $\phi_0(g) = a_0 g = g$. The inverse of $\phi_i$ is $\phi_j$, where $a_i^{-1} = a_j$ (this is easy to check).

We show that $G$ is "the same" group as the group $\Phi$, more precisely: The mapping $\lambda : G \to \Phi$, given by $\lambda(a_i) = \phi_i$, is an isomorphism. It is injective, because if $\lambda(a_i) = \lambda(a_j)$, then $\phi_i = \phi_j$ and thus $a_i = a_j$, since $\phi_i(a_0) = \phi_j(a_0)$. An injective mapping between two finite sets of the same size is surjective and thus bijective. We still have to show the homomorphism property of Definition 3.25 on page 44:

$$\lambda(a_i)\lambda(a_k) = \phi_i \phi_k = \phi_t = \lambda(a_t) = \lambda(a_i a_k), \text{ if } a_i a_k = a_t.$$

□

To prove properties of given groups, Cayley's theorem is not particularly useful, because it embeds a group of order $n$ in a group of order $n!$, which is much too large to be of practical use. For GAP, the theorem of course has great significance, because it implies that every finite group can be written as a group of permutations.

If we write a permutation as a product of disjoint cycles, the number of cycles that appear is unique, where we now and in the following include cycles of length 1: For example, $(1, 4, 9, 2)(3, 8, 5)(6)(7)$ as a product of disjoint cycles can only be written differently by swapping cycles or cyclically changing the order in a cycle, such as $(6)(9, 2, 1, 4)(7)(5, 3, 8)$. The number of cycles remains the same.

**Lemma 4.4** *Let $p \in S_n$ be any permutation and $t = (i, j)$ be a transposition. The number of cycles of $p$ and the number of cycles of $p \circ t$ differ by 1.*

**Proof**
**Case 1:** $i$ and $j$ are from different cycles of $p$:
$p$ then has, after possibly changing the order of the cycles, the following form:

$$p = (u, \ldots, i, a, \ldots, v)(x, \ldots, j, b, \ldots, y)\mu_3\mu_4\ldots\mu_r,$$

where $\mu_3, \mu_4, \ldots, \mu_r$ are further cycles. It follows that

$$p \circ t = (u, \ldots, i, a, \ldots, v)(x, \ldots, j, b, \ldots, y)\mu_3\mu_4\ldots\mu_r \circ (i, j)$$
$$= (u, \ldots, i, b, \ldots, y, x, \ldots, j, a, \ldots, v)\mu_3\mu_4\ldots\mu_r.$$

$p \circ t$ thus has one cycle less than $p$, and the assertion is shown for the first case.
**Case 2:** $i$ and $j$ are from the same cycle of $p$:
$p$ then has, after possibly changing the order of the cycles, the following form:

$$p = (u, \ldots, i, a, \ldots, j, b, \ldots, y)\mu_2\mu_3\ldots\mu_r,$$

where $\mu_2, \mu_3, \ldots, \mu_r$ are further cycles. It follows that

$$p \circ t = (u, \ldots, i, a, \ldots, j, b, \ldots, y)\mu_2\mu_3\ldots\mu_r \circ (i, j)$$
$$= (u, \ldots, i, b, \ldots, y)(j, a, \ldots)\mu_2\mu_3\ldots\mu_r.$$

$p \circ t$ therefore has one more cycle than $p$, and the assertion is also shown for the second case.                                                                                    □

**Theorem 4.5** *Every permutation is always either the product of an even or an odd number of transpositions. This means no permutation can be written in two different ways as a product of transpositions such that one product consists of an even and the other product of an odd number of transpositions.*

**Proof** We say that two integers have the same *parity* if either both are even or both are odd. Let $p \in S_n$. According to Theorem 4.2 we can write $p$ as a product of transpositions $p = t_1 t_2 \ldots t_m$. Let $r$ be the number of cycles of $p$.

We will show that the numbers $m$ and $n - r$ have the same parity. Since $n - r$ has nothing to do with the way we represent $p$ through transpositions, the assertion follows.

We prove the theorem by induction over $m$. If $m = 0$, then $p = id$ with $r = n$ cycles, and $m$ and $n - r$ have the same parity.

If $m = 1$, then $p = t_1$, and $p$ has $r = n - 1$ cycles (only 2 of the $n$ numbers from $T_n$ appear in the same cycle). So $n - r = 1$, and $m$ and $n - r$ have the same parity.

Inductively, we now assume that $m$ and $n - r$ have the same parity for all $m \leq k$. Let $p = t_1 t_2 \ldots t_{k+1}$. Let $r'$ be the number of cycles of $p' = t_1 t_2 \ldots t_k$. By the induction assumption, $k$ and $n - r'$ have the same parity. According to Lemma 4.4, the number of cycles $r$ of $p = p' t_{k+1}$ and $r'$ of $p'$ differ by 1. Therefore, $k + 1$ and $n - r$ have the same parity, just the opposite parity of $k$ and $n - r'$. □

**Definition 4.6** A permutation is called *even* if it is composed of an even number of transpositions and *odd* otherwise.

If we compose two even permutations, the result is again an even permutation. The identical permutation () is even, and therefore the inverse of an even permutation is also an even permutation. We see:

**Theorem 4.7** *The subset of even permutations of the group $S_n$ forms a subgroup, the* alternating group $A_n$.

The subgroup $A_n$ has index 2 in the group $S_n$. We claim that exactly half of all permutations of $S_n$ are even. Let $p \in S_n$ be a transposition. According to item 3 of Theorem 2.17 on page 24, we get all elements of $S_n$ by composing all elements of $S_n$ with $p$. However, each even permutation turns into an odd one and vice versa. So there are exactly as many even as odd permutations. According to Exercise 3. of Sect. 3.4, $A_n$ is normal in $S_n$.

We consider Example 2.22 again. Here we have realized the group $S_4$ geometrically as the symmetry group of the tetrahedron. Which isometries are even, and hence belong to $A_4$? According to the proof of Theorem 4.2, a cycle of length $m$ corresponds to an even permutation exactly when $m$ is odd. Thus the isometries

$$S_{4,1} = \{id, (2, 3, 4), (2, 4, 3), (1, 3, 4), (1, 4, 3), (1, 2, 4), (1, 4, 2),$$

$$(1, 2, 3), (1, 3, 2)\}$$

and

$$S_{4,2} = \{(2, 3)(1, 4), (1, 2)(3, 4), (1, 3)(2, 4)\}$$

are even permutations. These permutations correspond exactly to the rotations of the tetrahedron (see Example 2.22). Therefore, for the symmetry group $G \cong S_4$ of the tetrahedron, $G^+ \cong A_4$. The group $A_4$ is isomorphic to the subgroup of orientation-preserving isometries of the tetrahedron.

**Theorem 4.8** *The group $A_n$ is generated by all 3-cycles.*

*Proof* A 3-cycle $(i, j, k)$ can be written as a product of two transpositions:

$$(i, j, k) = (i, j) \circ (j, k). \tag{4.3}$$

So 3-cycles are included in $A_n$.

Every permutation from $A_n$ can be written as a product of an even number of transpositions. One decomposes such a product into pairs of adjacent transpositions. If such a pair has a common element, it can be written as a 3-cycle as in Eq. (4.3). If such a pair has no common element as in $(i, j)(k, l)$, one can write the identity $(j, k)(j, k)$ in between them and obtain two pairs of transpositions, each of which can be written as a 3-cycle. □

### Exercises

1. Prove that $A_3 \cong D_3^+$.
2. An *n-simplex* is a figure $\sigma^n \in \mathbb{R}^n$ consisting of the *convex hull* of $n + 1$ vertices which are mutually the same distance from each other, i.e., for any two points in $\sigma^n$, the line connecting these points is also in $\sigma^n$. Thus, a 2-simplex is an equilateral triangle and a 3-simplex is a regular tetrahedron. Prove that the symmetry group of the $n$-simplex is the group $S_{n+1}$.
3. Prove that the group $S_n$ is generated by $\{(1, 2), (2, 3), \ldots, (n - 1, n)\}$.

## 4.2  Operations of Groups on Sets

Let $G$ be the symmetry group of a figure $X$. If $x \in X$ is a point or an edge, then each isometry $g \in G$ maps the element $x$ to an element $g(x) \in X$. We say $G$ *operates* on $X$. The symmetry of the figure $X$ is described by the group $G$. We extend the concept of symmetry and consider more generally any sets $X$ (e.g. $T_n$), which do not necessarily have to be given by geometry. We also let groups operate on such general sets:

**Definition 4.9** Let $G$ be a group and $X$ a set. An *operation* of $G$ on $X$ is a mapping that assigns to each $g \in G$ and $x \in X$ an element $g(x) \in X$, such that

1. $\forall x \in X$ $e(x) = x$ (where $e$ denotes the identity element of $G$),
2. $\forall x \in X$ and $\forall g, h \in G$ $gh(x) = g(h(x))$ (associativity).

We say $G$ *operates* or *acts* on $X$ or $X$ is a $G$-set.

An operation is thus a mapping $\phi \colon G \times X \to X$ with the above two properties.

This operation is sometimes also called a *left operation*, in contrast to a *right operation*, where condition 2 changes to:

$$\forall x \in X \text{ and } \forall g, h \in G \; gh(x) = h(g(x)).$$

**Fig. 4.1** Dodecahedron

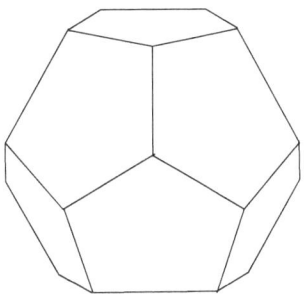

In the case of the symmetry group $G$ of a figure $F$ in the plane (or of a solid in space) these conditions are naturally fulfilled. The group $G$ operates on $F$. The elements of symmetry groups are mappings, and for mappings, the associative law 2 always applies. In addition, the identical mapping behaves as required in 1.

If $X$ is a $G$-set, then every $g \in G$ permutes the elements of $X$. The above mapping $\phi$, restricted to a fixed group element $g$, is a bijection. If $g(x) = g(y)$ for $x, y \in X$ and $g \in G$, then from the associativity of the operation $g^{-1}g(x) = g^{-1}g(y)$ and thus $x = y$. The mapping $\phi$ is therefore injective. It is also surjective: A given element $x \in X$ has $(g, g^{-1}(x))$ as its preimage. We can therefore also take the following as the definition of an operation of a group on a set:

> An operation of a group $G$ on a set $X$ is a homomorphism from $G$ to the symmetric group $S_X$ over $X$.

Let $G_{(5,3)}{}^+$ be the group of orientation-preserving symmetries of the dodecahedron of Fig. 4.1 (we will explain the name later).

The dodecahedron has 12 faces, each of which are regular pentagons. Three of these pentagons meet at a vertex of the dodecahedron. $G_{(5,3)}{}^+$ operates on the set of faces of the dodecahedron. Any rotation performed on the dodecahedron that transforms the dodecahedron into itself transforms any given face into another (generally different) face.

On page 33 we defined the stabilizer of a subset $S$ of a figure $F$ as the group of elements that leave $S$ invariant (i.e., map $S$ onto itself). Let $s$ be a face of the dodecahedron. Its stabilizer $G_{(5,3)}{}^+(s)$ is isomorphic to the group $D_5^+$, the rotations of the regular pentagon. Any rotation about the axis perpendicular to the center of $s$ by an angle $n * 360/5$ degrees ($n \in \{0, 1, 2, 3, 4\}$) not only transforms the dodecahedron into itself (is thus an element of $G_{(5,3)}{}^+$), but even $s$ into itself and is thus an element of the stabilizer $G_{(5,3)}{}^+(s)$.

The concept of the stabilizer can easily be extended to operations on sets: Assume the group $G$ acts on the set $X$. The *stabilizer* $G(x)$ of an element $x \in X$ is the set of group elements that leave $x$ invariant, i.e.

$$G(x) = \{g \in G \mid g(x) = x\}.$$

In the previous chapter, we have already seen that $G(x) < G$.

**Example 4.10** Let $X$ be the set of all quadrilaterals of the plane. $X$ is an $\mathcal{E}$-set, because every quadrilateral from $X$ is mapped onto a quadrilateral from $X$ by an isometry from $\mathcal{E}$. Let $Q \in X$ be any square in the plane. Then its stabilizer $\mathcal{E}(Q) = D_4$ consists exactly of the isometries that map $Q$ onto itself.

The set $X$ can be naturally decomposed with respect to $\mathcal{E}$: For any two congruent quadrilaterals $V, V' \in X$ there is an isometry $g \in \mathcal{E}$ with $g(V) = V'$. On the other hand, non-congruent quadrilaterals can never be mapped onto each other. For every quadrilateral $V \in X$ there is therefore the class $\mathcal{E}V$ of quadrilaterals congruent to $V$. $\mathcal{E}V$ consists precisely of the quadrilaterals that can be obtained from $V$ through an isometry. This class is called the *orbit* of $V$. Each class of congruent quadrilaterals forms an orbit, and the set of all quadrilaterals $X$ can be broken down into a union of disjoint orbits of congruent quadrilaterals.

**Definition 4.11** Let $X$ be a $G$-set and $x \in X$. The set

$$Gx = \{y \in X \mid \exists g \in G, \ y = g(x)\}$$

is called the *orbit* of $x$.

The orbit of an element $x \in X$ is therefore the set of all images of $x$ under the operation of elements of $G$.

**Example 4.12** Let $X = \{1, 2, 3, \ldots, 9\}$, and $G$ be the group generated by

$$g = (1, 5)(2, 7)(3, 4, 9, 8) \quad \text{and} \quad h = (6, 2)(4, 8, 5).$$

Then the orbits of the operation of $G$ on $X$ are as follows:

The orbit of 1 is $\{1, 5, 4, 8, 3, 9\}$ because $1 \xrightarrow{g} 5 \xrightarrow{h} 4 \xrightarrow{h} 8 \xrightarrow{g} 3$ and $4 \xrightarrow{g} 9$.

The orbit of 2 is $\{2, 7, 6\}$ because $2 \xrightarrow{g} 7$ and $2 \xrightarrow{h} 6$.

We consider the operation in GAP:

```
gap> G:=Group((1,5)(2,7)(3,4,9,8),
               (6,2)(4,8,5));;
gap> orbit1:=Orbit(G, 1 );
[ 1, 5, 4, 9, 8, 3 ]
gap> orbit2:=Orbit(G, 2 );
[ 2, 7, 6 ]
```

Two orbits are either disjoint or equal, because if the two orbits $Gx$ and $Gy$ have a common element $z \in X$ and $x' \in Gx$ and $y' \in Gy$ are arbitrary elements, then there is a $g \in G$ with $g(x') = z$, because $x', z \in Gx$. Similarly, there is an $h \in G$ with $h(y') = z$, because $y', z \in Gy$. This implies $h^{-1}g(x') = y'$, so $x'$ and $y'$ are in the same orbit. This leads to:

**Theorem 4.13** *Let $X$ be a $G$-set. Then $X$ is a disjoint union of orbits with respect to $G$.*

**Fig. 4.2**   Square in the plane

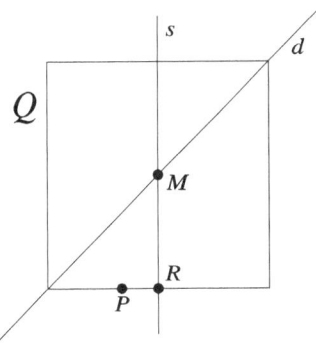

**Definition 4.14**  Let $X$ be a $G$-set. $x \in X$ is called a *fixed point* of $g \in G$ if $g(x) = x$. Let $X^g$ be the set of all fixed points of $G$, i.e.,

$$X^g = \{x \in X \mid g(x) = x\}.$$

**Example 4.15**  Let $Q$ be the square with side length 1 in $\mathbb{R}^2$ with vertices $(0, 0), (0, 1), (1, 0), (1, 1)$ (see Fig. 4.2). The symmetry group of $Q$ is the group $D_4$.

1. Let $t$ be the reflection on the diagonal $d$ through $(0, 0)$ and $(1, 1)$. The orbit of the vertex $(0, 0)$ consists of all 4 vertices of $Q$. The stabilizer of $(0, 0)$ is

$$D_4((0, 0)) = \{id, t\}.$$

   $X^t$ consists of all points of $d$.
2. The stabilizer of $P = (1/3, 0)$ is the trivial group. The orbit of $P$ consists of 8 points.
3. Let $M = (1/2, 1/2)$ be the center of $Q$. $D_4(M) = D_4$. The orbit of $M$ consists only of $M$ itself.
4. The orbit of $R = (1/2, 0)$ consists of 4 points. The stabilizer of $R$ is $\{id, s\}$, where $s$ is the reflection over the line through $(1/2, 0)$ and $(1/2, 1)$.

**Example 4.16**  The symmetric group $S_n$ operates on the set $T_n = \{1, \ldots, n\}$. The stabilizer $S_n(n)$ consists of all those permutations that map the number $n$ onto itself. Thus, $S_n(n) \cong S_{n-1}$. The orbit of $n$ in the group $S_n$ is the entire set $T_n$, because for each $i \in T_n$, $i \neq n$, there is a permutation, for example $(n, i)$, that maps $n$ onto $i$.

**Example 4.17**  Let $W$ be the group of the cube of Fig. 1.8. We inscribe a tetrahedron as in Fig. 3.3 on page 35. The cube group $W$ acts on the vertices of the cube. Each isometry of the cube maps the tetrahedron onto a tetrahedron, either onto itself or onto the second possible position of the tetrahedron in the cube. The vertices of the tetrahedron in the cube are the vertices 1, 3, 6, 8, and the cube group operates on the set:

$$\{\{1, 3, 6, 8\}, \{2, 4, 5, 7\}\}.$$

We generate in GAP the cube group W from the generators a, b, c and consider the orbit of $\{1, 3, 6, 8\}$.

```
gap> a:=(1,2)(5,6)(4,3)(8,7);;
gap> b:=(1,3)(5,7);; c:=(5,4)(6,3);;
gap> W:=Group(a,b,c);;
gap> bahn := Orbit( W, [1,3,6,8], OnSets );
[ [ 1, 3, 6, 8 ], [ 2, 4, 5, 7 ] ]
gap> Tet := Stabilizer( W, [1,3,6,8], OnSets );
Group([ (3,6)(4,5), (2,4,5)(3,8,6), (1,6)(2,5)(3,8)(4,7),
    (1,8)(2,7)(3,6)(4,5) ])
gap> Order( Tet );
24
gap> IsomorphismGroups( Tet, SymmetricGroup(4) );
[ (3,6)(4,5), (2,4,5)(3,8,6), (1,6)(2,5)(3,8)(4,7),
    (1,8)(2,7)(3,6)(4,5) ] ->
[ (1,2), (1,4,2), (1,4)(2,3), (1,2)(3,4) ]
```

The orbit of the tetrahedron consists of only two elements, both of which we can "see" in the cube. The stabilizer of the tetrahedron is a subgroup with 24 elements and is isomorphic to the group $S_4$. This means: Every isometry of the tetrahedron is also one of the cube.

Let $S$ be the set of faces of the dodecahedron. We already know that $G_{(5,3)}^+$ operates on $S$. Let $s \in S$ be a face. For any other face $s' \in S$ there is an isometry $g \in G_{(5,3)}^+$ such that $g(s) = s'$. We can rotate any pentagon to any other. This means, however, that the orbit of $s$ completely encompasses $S$, so $G_{(5,3)}^+ s = S$. Therefore, with respect to the operation of $G_{(5,3)}^+$, $S$ consists of one orbit only. In this case, we speak of a *transitive* operation.

**Definition 4.18** Let $X$ be a $G$-set. If there is only one orbit with respect to the operation of $G$ on $X$, then the operation is called *transitive*.

The following is obviously equivalent to this definition: $G$ acts transitively on $X$ if, for any $x \in X$, $Gx = X$.

We easily check the transitivity in GAP: We consider the group of the tetrahedron, which we have already studied in the last chapter with GAP:

```
gap> Tetra:=Group((2,4),(1,2),(1,3),(1,4),(2,3),(3,4));
Group([ (2,4), (1,2), (1,3), (1,4), (2,3), (3,4) ])
gap> Orbit(Tetra,1);
[ 1, 4, 2, 3 ]
```

Orbit(Tetra,1) gives us the orbit of 1 in the group Tetra. It encompasses all four vertices of the tetrahedron. Therefore, the group of the tetrahedron operates transitively on the vertices of the tetrahedron.

If $X$ is the set of quadrilaterals in the plane of Example 4.10, then $\mathcal{E}$ does not operate transitively. There are many orbits, one for each class of congruent quadrilaterals.

We consider the stabilizer $\mathcal{E}(0)$ of the origin in the symmetry group of the plane. It consists of all those isometries of the plane which map the origin onto itself. According to Theorem 1.6 on page 9, these are reflections over lines through the origin and rotations about the origin. For all these isometries, every circle about the origin remains invariant, i.e., $\mathcal{E}(0)$ is isomorphic to the symmetry group of the circle. This group is called the *orthogonal group* of the plane, abbreviated $O_2$, and was already introduced in Definition 3.5 on page 33 (compare with Exercise 3. of Sect. 3.3). The lower index 2 stands for the dimension of the plane. The stabilizer of the origin in $\mathbb{R}^3$ is the orthogonal group $O_3$ and consists of all isometries which map a sphere (all points at distance 1 from the origin in $\mathbb{R}^3$) onto itself.

**Theorem 4.19** $O_3^+$ *consists of all rotations about lines through the origin in $\mathbb{R}^3$.*

A proof can be found in the appendix in the section on matrices.

Obviously, such an orthogonal group exists in every dimension. The stabilizer of the origin in $\mathbb{R}^n$ is the group $O_n$. This group acts transitively on the points of the *sphere* of dimension $n - 1$, the so-called $n - 1$-*sphere*, i.e., all points at distance 1 from the origin in $\mathbb{R}^n$.

$\mathcal{E}$ operates transitively on the set of points of the plane $\mathbb{R}^2$, because there is only one orbit with respect to this operation. For any two points $x, y \in \mathbb{R}^2$ there is a translation $\tau \in \mathcal{E}$ with $\tau(x) = y$.

An important example of a transitive operation is the following:

**Example 4.20** Let $(G, \cdot)$ be any group. Then $G$ acts on itself (more precisely: on the set of its elements) through the group operation $g(h) = g \cdot h$.

This operation is transitive, because there is only one orbit. Let $x \in G$ be any group element. Then the orbit of $x$ encompasses the entire group $G$, because for any $y \in G$ there exists a $g \in G$ such that $y = gx$, and thus $y$ is in the orbit of $x$.

**Example 4.21** Let $H$ be a subgroup of a group $(G, \cdot)$. Then $H$ acts on $G$ by $h(g) = h \cdot g$, where $h \in H$ and $g \in G$. The orbit of an element $g \in G$ consists of all elements of the form $h \cdot g$ with $h \in H$, so it is equal to the right coset $Hg$.

The notation for an orbit coincides with the notation for a right coset. In fact, this is the reason for the notation for orbits. Similarly, we obtain the left cosets if we define $h(g) = g \cdot h$.

### Exercises

1. Let $W$ be a cube in $\mathbb{R}^3$ and $G$ its symmetry group. Calculate the stabilizer of

   (a) any vertex,
   (b) any edge,
   (c) any face,
   (d) two opposite faces,

(e)  two opposite edges,
(f)  two opposite edges and one face

of the cube in $G$.

2. Check whether the following statements are true or false. Justify!

(a)  The group $D_4$ operates on the edges of a regular hexagon.
(b)  The group $D_4$ operates on the edges of a regular octagon.
(c)  The group $D_7$ operates on the vertices of a regular 56-gon.
(d)  The group $D_4$ operates on any three edges of a square.
(e)  The group $D_3$ operates on the faces of a tetrahedron.

3. Let $W$ be the symmetry group of a cube. Show that the subgroup of orientation-preserving isometries $W^+$ is isomorphic to the group $S_4$. (Hint: Permute the 4 main diagonals of the cube with rotations, read Sect. 1.4 again and use GAP.)
4. Let $W^+$ be the subgroup of orientation-preserving isometries of the symmetry group of a cube. Is the operation of $W^+$ on the vertices (edges, faces) of the cube transitive?

## 4.3  Conjugation

We have already defined conjugation in Sect. 3.3 as follows:

**Definition 4.22**  Let $G$ be a group and $g, h \in G$. The element

$$g^h = h^{-1}gh$$

is called *conjugation* of $g$ with $h$.

This conjugation is a special operation of a group on itself, if we understand it as a right operation. For the left operation, one must take the mapping $(h, g) \to hgh^{-1}$. We mostly work with the right operation.

One can immediately see that $g^h = g$ holds exactly when $g$ and $h$ *commute*, i.e., when $gh = hg$ holds in the group $G$.

If $g, h, j, k$ are elements of a group $G$, then $(gj)^h = g^h j^h$ and $g^{hk} = (g^h)^k$, because

$$(gj)^h = h^{-1}gjh = h^{-1}ghh^{-1}jh = g^h j^h$$

and

$$g^{hk} = (hk)^{-1}ghk = k^{-1}h^{-1}ghk = (g^h)^k.$$

So, laws analogous to the power laws apply, hence the notation.

**Definition 4.23** The elements $g, j \in G$ are called *conjugate to each other* if there is an $h \in G$ such that $g^h = j$.

It is easy to see when two permutations $g, h \in S_n$ are conjugate to each other. They are conjugate exactly when they have the same *cycle structure*, i.e., the cycles in the respective cycle notation have the same lengths. For example, $g = (6, 9)(1, 3, 4)(2, 5, 7, 8)$ and $h = (1, 2)(3, 4, 5)(6, 7, 8, 9)$ are conjugate in $S_9$. They are each disjoint products of cycles of lengths $2, 3$ and $4$. We prove this in general:

**Theorem 4.24** *Let $g, p \in S_n$ be two permutations. Then $g$ and $p$ are conjugate to each other if and only if the following statement holds: If $g$ is a disjoint product of cycles of lengths $k_1, \ldots, k_m$, then $p$ is also a disjoint product of cycles of lengths $k_1, \ldots, k_m$.*

**Proof** Let $g, p$ be conjugate. So there is an $h \in S_n$ with $g = h^{-1}ph$, hence $p = hgh^{-1}$. Let $g(i) \in \{1, \ldots, n\}$ be the image of $i$ under the permutation $g$. If $g(i) = j$, then

$$(hgh^{-1})h(i) = hg(i) = h(j).$$

Therefore, if $g$ sends the number $i$ to $j$, then $p = hgh^{-1}$ sends the number $h(i)$ to $h(j)$. So $g$ and $hgh^{-1}$ have the same cycle structure.

For the converse: If $g, g'$ have the same cycle structure, then one can find an $h$ such that $g' = hgh^{-1}$, by simply mapping corresponding elements in the cycles of $g, g'$ onto each other with $h$.

For example, if $g = (3, 6, 9) \ldots$ and $g' = (2, 4, 1) \ldots$, then define $h(3) = 2$, $h(6) = 4$ and $h(9) = 1$. Then for example $g'(2) = hgh^{-1}(2) = hg(3) = h(6) = 4$, as desired. $\square$

"Being conjugate" is an equivalence relation. Every element is conjugate to itself (reflexive), as one can see by conjugating with the identity element. By conjugating with the inverse we see that if $g$ is conjugate to $j$, then $j$ is also conjugate to $g$. "Being conjugate" is therefore symmetric. Transitivity is proven in Exercise 1. at the end of this section.

The set of equivalence classes of $G$ under the equivalence relation given by conjugation is denoted $G_*$.

**Definition 4.25** For a finite group $G$, the function ord: $G \to \mathbb{N} \cup \{\infty\}$, which assigns to each element its order, is the so-called *order function*.

The polynomial

$$p_G(t) = \sum_{g \in G_*} t^{\text{ord}(g)}$$

is called the *generating polynomial* of the order function of $G$.

One can easily see that if two elements are conjugate, then they have the same order. If $h^{-1}gh = j$ and $n \in \mathbb{N}$, then

$$j^n = (h^{-1}gh)^n = h^{-1}ghh^{-1}gh \ldots h^{-1}gh = h^{-1}g^nh.$$

Therefore, $j^n = 1$ exactly when $g^n = 1$, and the generating polynomial makes sense.

**Example 4.26** The generating polynomial for the group $S_8$ is

$$t^{15} + t^{12} + t^{10} + t^8 + t^7 + 5t^6 + t^5 + 4t^4 + 2t^3 + 4t^2 + t.$$

For example, the term $4t^4$ is due to the 4 cycle structures:

$$(1, 2, 3, 4);$$
$$(1, 2, 3, 4)(5, 6);$$
$$(1, 2, 3, 4)(5, 6, 7, 8);$$
$$(1, 2, 3, 4)(5, 6)(7, 8).$$

For $g \in G$, the *conjugacy class* of $g$ in $G$

$$Kg = \{hgh^{-1} \mid h \in G\}$$

is the set of all elements in $G$ conjugate to $g$. Because "being conjugate" is an equivalence relation, we have

$$G = \bigcup_{g \in G_*} Kg.$$

**Theorem 4.27** *Let $X$ be a $G$-set. If $g$ and $h$ are conjugate in $G$, then they have the same number of fixed points.*

**Proof** Let $h = jgj^{-1}$ and $x \in X$ be a fixed point of $g$ (see Definition 4.14). Then

$$h(j(x)) = jgj^{-1}(j(x)) = jg(x) = j(x).$$

$j(x)$ is therefore a fixed point of $h$ and $j$ thus maps the fixed points of $g$ to those of $h$, i.e., the set $X^g$ to $X^h$.

Since $g = j^{-1}hj$, the same argument with $g$ and $h$ in swapped roles shows that $j^{-1}$ maps the set $X^h$ to $X^g$. Therefore, $j$ is a bijective mapping from $X^g$ to $X^h$, so these two fixed point sets must be of the same size. $\qquad\square$

For a subgroup $H < G$, let $H^g = \{g^{-1}hg \mid h \in H\}$ be a subgroup *conjugate to* $H$.

**Theorem 4.28** *If H is a subgroup of the group G and $g \in G$, then $H^g$ is a subgroup of G.*

**Proof** For two elements $g^{-1}hg, g^{-1}h'g \in H^g$ their product $g^{-1}hgg^{-1}h'g = g^{-1}hh'g$ is again in $H^g$, which shows closure. If $e \in G$ is the identity element, then $g^{-1}eg = e$ is in $H^g$. The inverse of $g^{-1}hg$ is $g^{-1}h^{-1}g \in H^g$. □

**Theorem 4.29** *Let G be a group, $H < G$ and $g \in G$. Then H is isomorphic to $H^g$.*

The proof is formulated as an exercise at the end of the section.

Being conjugate is also an equivalence relation among subgroups. Therefore, there are equivalence classes of conjugate subgroups.

**Theorem 4.30** *Let X be a G-set, and assume $x, y \in X$ are from the same orbit. Then the stabilizers of x and y are conjugate subgroups.*

**Proof** Let $g(x) = y$. We show that $g\, G(x)g^{-1} = G(y)$. Let $h \in G(x)$. Then

$$ghg^{-1}(y) = ghg^{-1}(g(x)) = gh(x) = g(x) = y.$$

It follows that $g\, G(x)g^{-1} \subset G(y)$. If we swap the roles of $x$ and $y$, we get $g^{-1}G(y)g \subset G(x)$ or $G(y) \subset g\, G(x)g^{-1}$. Altogether, $g\, G(x)g^{-1} = G(y)$. □

In GAP we consider the conjugacy classes of subgroups of the group $A_4$.

```
gap> C:=ConjugacyClassesSubgroups(AlternatingGroup(4));
[ Group( () )^G, Group( [ (1,2)(3,4) ] )^G,
  Group( [ (2,4,3) ] )^G,
  Group( [ (1,3)(2,4), (1,2)(3,4) ] )^G,
  Group( [ (1,3)(2,4), (1,2)(3,4), (2,4,3) ] )^G ]
```

In the conjugacy class of the second group, the group $\langle(1,2)(3,4)\rangle$ (isomorphic to $\mathbb{Z}_2$), there are three groups:

```
gap> Elements(C[2]);
[ Group([ (1,2)(3,4) ]), Group([ (1,3)(2,4) ]),
  Group([ (1,4)(2,3) ]) ]
```

We again look at the decomposition of the plane into squares of Example 3.18 on page 39. Its symmetry group is the group $G_{(4,4)}$ with subgroup $U < G_{(4,4)}$ isomorphic to the group $D_4$ of the square generated by the reflections $s_a, s_b$ of Fig. 4.3, as we have clarified in Sect. 3.1. We can conjugate $U$ with a translation $\mu$ (see Fig. 4.3), i.e., form $\mu^{-1} \circ U \circ \mu = \{\mu^{-1} \circ u \circ \mu \mid u \in U\}$. From Fig. 4.3 it is easy to see that we thereby obtain another group that is isomorphic to the group of the square. This time we obtain all isometries of the square labeled $A$ in Fig. 4.3 as a subgroup of the group $G_{(4,4)}$.

Conjugation of $U$ with all translations yields all reflections through all square centers and all rotations about these points. If one additionally conjugates the group $V$ generated by $s_b, s_c$ (which is also isomorphic to $D_4$) with all translations, one obtains all elements of $G_{(4,4)}$. It is not difficult to see that the horizontal and vertical translations $\mu$ and $\tau$ can be represented as products of $s_a, s_b, s_c$: One obtains a

**Fig. 4.3**  Generators of the group $G_{(4,4)}$

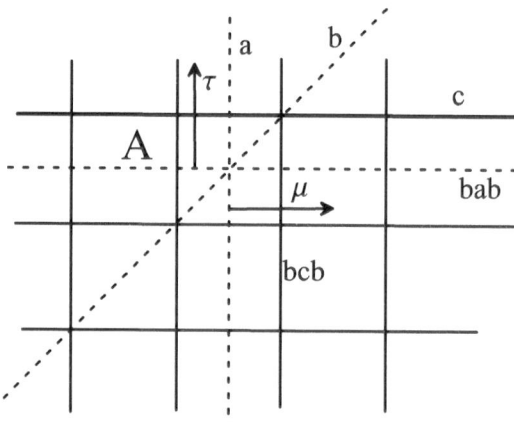

**Fig. 4.4**  Conjugation of a reflection with a translation

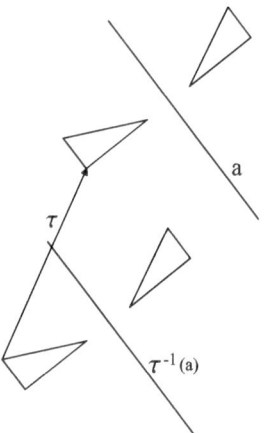

translation according to Theorem 1.9 by composing reflections along parallel axes. Thus we obtain: $\tau = s_c \circ s_b s_a s_b$ and $\mu = s_b s_c s_b \circ s_a$. Since every translation from $G_{(4,4)}$ can be obtained by performing $\tau$ and $\mu$ in succession, we have proven:

**Theorem 4.31**  *The group $G_{(4,4)}$ is generated by the reflections $s_a$, $s_b$, $s_c$ of Fig. 4.3.*

If $U$ is a subgroup of $\mathcal{E}$, then a subgroup conjugate to $U$ is isomorphic to $U$, but their elements are isometries described by lines/points in different locations of the plane. Two figures $F_1$, $F_2$ in the plane are called *congruent* if there is an isometry that maps $F_1$ onto $F_2$. Thus, if the figures $F_1$, $F_2$ are congruent, then their symmetry groups (as subgroups of the group of the plane) are conjugate.

**Example 4.32**  Let $\tau$ be a translation of the plane and $s_a$ the reflection over a straight line $a$. Then the conjugation $\tau^{-1} s_a \tau$ of $s_a$ with $\tau$ is the reflection over the straight line $\tau^{-1}(a)$ (the line $a$ shifted by $\tau^{-1}$), as can be seen in Fig. 4.4.

In the following theorem we generalize the phenomenon of Example 4.32:

**Theorem 4.33** *Let $\alpha \in \mathcal{E}$ be any isometry of the plane $\mathbb{R}^2$.*

1. *Let $\tau$ be a translation with invariant line $l$. Then $\alpha\tau\alpha^{-1}$ is a translation of the same length with invariant line $\alpha(l)$.*
2. *Let $d$ be a rotation about the point $P$. Then $\alpha d\alpha^{-1}$ is the rotation about $\alpha(P)$ by the same angle.*
3. *Let $s$ be the reflection over the line $l$. Then $\alpha s\alpha^{-1}$ is the reflection over the line $\alpha(l)$.*

*Proof*

1. We have $\tau(l) = l$. Then $\alpha\tau\alpha^{-1}(\alpha(l)) = \alpha(l)$. This holds for all lines $l$ invariant under $\tau$.
2. We have $d(P) = P$. Then $\alpha d\alpha^{-1}(\alpha(P)) = \alpha(P)$. Because $\alpha d\alpha^{-1}$ is orientation-preserving with fixed point $\alpha(P)$, the assertion follows.
3. We have $s(P) = P$ for all points $P \in l$. Then $\alpha s\alpha^{-1}(\alpha(P)) = \alpha(P)$. Thus, the points $\alpha(P)$ are fixed points of $\alpha s\alpha^{-1}$.

$\square$

**Exercises**

1. Prove that the relation "being conjugate" in a group $G$ is transitive. So prove: For all $g, h, j \in G$, if $g$ is conjugate to $h$ and $h$ is conjugate to $j$, then $g$ is conjugate to $j$.
2. Determine the generating polynomial for the group $S_7$.
3. Let $G$ be a group, $H < G$ and $g \in G$. Prove that $H$ is isomorphic to $H^g$. Specifically, prove that the mapping $f_g : H \to H^g$ defined by $f_g(h) = g^{-1}hg$ is a bijective mapping that satisfies the condition $f_g(h)f_g(h') = f_g(hh')$ for all $h, h' \in H$.
4. Let $\mathcal{E}$ be the symmetry group of the plane and $d$ a rotation by $360/n$ degrees about the origin. Let $H = \langle d \rangle$. Then $H < \mathcal{E}$. Determine the subgroups conjugate to $H$.
5. Let $W$ be the group of the cube and $S$ a face of the cube. Determine the subgroups of $W$ that are conjugate subgroups of the stabilizer $W(S)$.
6. Determine the conjugate subgroups to the stabilizers of Example 4.15.

## 4.4   The Orbit-Stabilizer Theorem and the Class Equation

We once again consider the group of orientation-preserving isometries of the dodecahedron $G_{(5,3)}^+$ (see Fig. 4.1). Let $S$ be the set of faces of the dodecahedron and $s \in S$. Its stabilizer $G_{(5,3)}^+(s)$ is isomorphic to the group $D_5^+$, consisting of the rotations of the regular pentagon, as we have seen above. We have also seen above that $G_{(5,3)}^+s = S$ holds. Any group element $g \in G_{(5,3)}^+$ maps the face $s$ to a (generally different) face $s'$. We can associate the coset $g \cdot G_{(5,3)}^+(s)$ with the orbit element $g(s) = s'$. We show in Theorem 4.34 that this assignment is even bijective.

The cosets $g \cdot G_{(5,3)}{}^+(s)$ correspond to the stabilizers of the $s_j \in S$. These are in bijective relation to the $s_j \in S$, thus the orbit of $s$.

**Theorem 4.34** *Let X be a G-set and $x \in X$. Then the map $\phi \colon G/G(x) \to Gx$ defined by $\phi(gG(x)) = g(x)$ is bijective.*

**Proof** To show that $\phi$ is a well-defined mapping, we have to show that $gG(x) = hG(x)$ implies $g(x) = h(x)$. If $gG(x) = hG(x)$, there is a $t \in G(x)$ with $h = gt$. Since $t$ leaves the element $x$ fixed, it follows that $h(x) = gt(x) = g(x)$. The mapping is therefore well-defined.

The mapping $\phi$ is surjective, because for every $g(x)$ there is a coset $gG(x)$. $\phi$ is also injective: If the cosets $gG(x)$ and $hG(x)$ have the same image $g(x) = h(x)$, then $x = g^{-1}h(x)$. Thus, $g^{-1}h$ is an element of the stabilizer $G(x)$, and $g$ and $h$ differ only by an element of this stabilizer. It follows that $gG(x) = hG(x)$.      □

$G_{(5,3)}{}^+s$ is the orbit of the face $s$. $|G_{(5,3)}{}^+s|$ is the *length* of the orbit, i.e., the number of its elements, in our case 12. The stabilizer $G_{(5,3)}{}^+(s)$ consists of all rotations of a regular pentagon in the dodecahedron, so $|G_{(5,3)}{}^+(s)| = 5$. In total, $|G_{(5,3)}{}^+| = |G_{(5,3)}{}^+(s)| \cdot |G_{(5,3)}{}^+s|$. The group of orientation-preserving isometries of the dodecahedron therefore has order 60. Because $G_{(5,3)}(s) = D_5$ and $|D_5| = 10$, it follows from $|G_{(5,3)}| = |G_{(5,3)}(s)| \cdot |G_{(5,3)}s|$ that the symmetry group $G_{(5,3)}$ of the dodecahedron has order 120. This is an application of *Lagrange's Theorem* (Theorem 3.14):

**Theorem 4.35 (Orbit-Stabilizer Theorem)** *Let X be a G-set, G a finite group and $x \in X$. Then*

$$|G| = |G(x)| \cdot |Gx|.$$

**Proof** $G(x)$ is a subgroup of $G$. Therefore, according to Corollary 3.16,

$$(\text{Order of } G) = (\text{Order of } G(x)) \cdot (\text{Number of cosets}).$$

According to Theorem 4.34, the number of cosets is equal to the length of the orbit.
                                                                                                     □

We apply the orbit-stabilizer theorem to Example 4.16:

$$n! = |S_n| = |S_n(n)| \cdot |S_n n| = |S_{n-1}| \cdot |T_n| = (n-1)! \cdot n.$$

Here, $T_n = \{1, 2, \ldots, n\}$. This corresponds to the recursive definition of the factorial function $n! = (n-1)! \cdot n$.

**Example 4.36** We reconsider Example 4.17 on page 67. The orbit of the tetrahedron $T$ inscribed in the cube consists of 2 elements and the stabilizer consists of 24. The orbit-stabilizer theorem gives the order of the cube group (which we already know):

$$|W| = |W(T)| \cdot |WT| = 24 \cdot 2 = 48.$$

If we consider the stabilizer $W(s)$ of a face $s$ of the cube, we obtain the same result, because 8 symmetries stabilize a face ($W(s) \cong D_4$) and the cube has 6 faces:

$$|W| = |W(s)| \cdot |Ws| = 8 \cdot 6 = 48.$$

Directly from the orbit-stabilizer theorem, we obtain:

**Corollary 4.37** *Assume the group $G$ acts on the set $X$. Let $x \in X$ with trivial stabilizer, i.e., $G(x) = \{e\}$. Then there is a one-to-one correspondence between the elements of $G$ and the elements of the orbit of $X$.*

We have already seen in Sect. 4.3 that a group $G$ operates on itself by conjugation, i.e. $g(x) = gxg^{-1}$ for $g, x \in G$. The stabilizer of an element $x \in G$ with respect to conjugation is called the *centralizer* of $x$ and is abbreviated as $Z(x)$. These are all the elements $g \in G$ for which $gxg^{-1} = x$ holds, or expressed differently:

$$Z(x) = \{g \in G \mid gx = xg\}.$$

The centralizer of $x$ consists of the elements that commute with $x$. Every element commutes with itself, so $x \in Z(x)$. The identity element of a group commutes with all group elements, so $Z(id) = G$. The centralizer is a subgroup because it is a stabilizer. If $G$ is an abelian group and $g \in G$, then $Z(g) = G$.

In GAP, we determine the centralizer of a rotation $d$ by $90°$ in the group $D_4$. It turns out that $Z(d) = D_4^+$.

```
gap> D4:=Group((1,2)(3,4),(1,2,3,4));;
gap> Elements(Centralizer(D4,(1,2,3,4)));
[ (), (1,2,3,4), (1,3)(2,4), (1,4,3,2) ]
```

We have already defined conjugacy classes in the previous section. Here comes an alternative definition, which of course leads to the same result: The orbit of an element $x \in G$ under the operation of conjugation is called the *conjugacy class* of $x$,

$$Kx = \{h \in G \mid \exists g \in G, h = gxg^{-1}\}.$$

According to the orbit-stabilizer theorem, for finite groups $G$ and for all elements $x \in G$,

$$|G| = |Kx| \cdot |Z(x)|.$$

Since conjugacy classes are orbits, according to Theorem 4.13 they decompose the group into disjoint classes. So in the case of a finite group, we have proven the so-called *class equation*:

**Theorem 4.38** *Let G be a finite group with conjugacy classes $K_1, \ldots, K_n$. Then the following holds:*

$$|G| = |K_1| + |K_2| + \cdots + |K_n|. \tag{4.4}$$

Here one has to be careful with the notation. The conjugacy classes are simply enumerated by $K_i$, whereas $Kx$ is the conjugacy class of the element $x \in G$. The identity always forms its own conjugacy class and to avoid naming problems, it should be denoted by $K_1$.

**Example 4.39** The conjugacy classes of the group $D_3$ are:

$$\{id\}, \quad \{d, d^2\}, \quad \{s, sd, sd^2\},$$

where $s \in D_3$ is any reflection and $d \in D_3$ is the rotation by $120°$.

It holds that $sds = d^2$, as can be seen in Fig. 1.1, and therefore $d$ and $d^2$ are conjugate to each other (remember: $s^{-1} = s$). We have (because $s^2 = 1$) $s\, sd\, s = ds = sd^2$, and therefore $sd$ and $sd^2$ are conjugate. Finally,

$$(sd)\, s\, (d^{-1}s) = d^2\, d^{-1}s = ds = sd^2,$$

which shows that $s$ is conjugate to $sd^2$. Now we still have to show that two elements from different classes are not conjugate to each other. This follows from the insight of p. 72, that conjugate elements have the same order, and the observation that elements of different classes in our example have different orders.

In GAP we determine the elements of the conjugacy classes of the group $D_3$.

```
gap> D3:=Group((1,2),(1,2,3));;
gap> cl:=ConjugacyClasses(D3);
[ ()^G, (1,3,2)^G, (2,3)^G ]
gap> Elements(cl[1]);
[ () ]
gap> Elements(cl[2]);
[ (1,2,3), (1,3,2) ]
gap> Elements(cl[3]);
[ (2,3), (1,2), (1,3) ]
```

According to the orbit-stabilizer theorem (Theorem 4.35), the length of each orbit is a divisor of the group order and, since conjugacy classes are orbits, each summand on the right-hand side of the class equation (4.4) divides the group order.

In GAP we also determine the numbers of elements of the conjugacy classes of $A_5$.

```
gap> A5:=AlternatingGroup(5);;
gap> cl:=ConjugacyClasses(A5);
[ ()^G, (1,2)(3,4)^G, (1,2,3)^G,
  (1,2,3,4,5)^G, (1,2,3,5,4)^G ]
gap> List(cl, i -> Order(i));
[ 1, 15, 20, 12, 12 ]
```

The class equation of the group $A_5$ is therefore

$$|A_5| = 1 + 15 + 20 + 12 + 12. \tag{4.5}$$

The 15 comes from elements of order 2, the 20 from elements of order 3 and the two 12s from elements of order 5, as can be seen from the output of GAP.

**Definition 4.40** A nontrivial group is called *simple* if it contains no normal subgroups other than the trivial group and itself.

If $p$ is a prime number, then $\mathbb{Z}_p$ is simple. Every subgroup must divide the group order according to Lagrange's theorem, and therefore $\mathbb{Z}_p$ only has the trivial group and itself as subgroups. All other groups have proper subgroups, as easily follows from the first Sylow theorem (Theorem 7.2 on page 129). Whether these subgroups are normal is not clear, however.

For several years now, all finite simple groups have been known. Classifying them had been a goal of group theory for a long time. Here *classifying* means creating a list of all finite simple groups such that each finite simple group is isomorphic to exactly one group on this list. The entire proof comprises numerous articles whose combined length runs to several thousand pages (see for example [Bog08]).

The following theorem plays a central role in Galois theory. With its help, it can be shown that the roots of fifth-degree polynomials generally cannot be found by a formula.

**Theorem 4.41** *The alternating group $A_5$ is a simple group.*

***Proof*** Let $N \lhd A_5$, and assume $N$ is not the trivial group. Then there is an element $x \neq 1$ in $N$. But then, by definition of normal subgroup, the whole conjugation class $Kx$ must lie in $N$. $N$ consists therefore of a union of conjugation classes. The order of $N$ must therefore be the sum of some numbers on the right-hand side of (4.5), where the 1 must be included, as every subgroup contains the identity element. On the other hand, the order of $N$ must be a divisor of 60. Both conditions can only be met if all numbers on the right-hand side are taken, i.e. $N = A_5$. Therefore, $A_5$ contains no proper normal subgroups. $\qquad\square$

We recall the definition of the center from Exercise 1. of Sect. 3.5: The elements of a group $G$ that commute with all group elements form a subgroup, the *center* of $G$:

$$C(G) = \{h \in G \mid \forall g \in G \ gh = hg\}.$$

It is easy to see that each center element forms its own conjugation class. If $x$ is in the center of $G$, then $gxg^{-1} = x$ follows for all group elements $g \in G$. The identity element is always in the center of a group, and thus at least one 1 always appears on the right-hand side of (4.4).

**Theorem 4.42** *Let* Inn$(G)$ *be the group of inner automorphisms of a group G. Then*

$$\text{Inn}(G) \cong G/C(G).$$

***Proof*** Consider the homomorphism $\phi: G \rightarrow \text{Aut}(G)$ which maps each group element $g \in G$ to the inner automorphism $h \rightarrow ghg^{-1}$. The image consists of all inner automorphisms, and the kernel is exactly

$$\{g \in G \mid h = ghg^{-1}, \forall h \in G\} = \{g \in G \mid hg = gh, \forall h \in G\} = C(G).$$

The result follows from the first isomorphism theorem.                          □

**Definition 4.43** A *p-group* is a group whose order is a power of a prime number $p$.

**Theorem 4.44** *The center of a p-group is nontrivial (i.e. contains more elements than just the identity element).*

***Proof*** Let $|G| = p^k$, where $k > 0$ and $p$ is a prime number. The left-hand side of (4.4) is thus $p^k$. According to the orbit-stabilizer theorem, each summand on the right-hand side of (4.4) must be a divisor of $p^k$, thus is itself a $p$-power. The elements of the center are exactly those that induce a 1 on the right-hand side of the class equation. If only the identity element were in the center, then all other conjugation classes would have proper $p$-powers as the number of elements. It would thus follow from the class equation that

$$p^k = 1 + \sum (\text{multiples of } p).$$

This is impossible for $k > 0$ because the left-hand side is divisible by $p$, but the right-hand side is not.                          □

  The group $D_4$ is a $p$-group due to $|D_4| = 2^3$ (more precisely: a 2-group). We want to let GAP calculate the size of its center. Many groups are predefined in GAP. The group $D_4$ is DihedralGroup(8) in GAP.

Centre gives us the center.      `gap> Order(Centre(DihedralGroup(8)));`
                                 `2`

  Apart from the identity, the rotation by 180° is also in the center of the group $D_4$ (compare Exercise 1. of Sect. 3.5).

**Theorem 4.45** *The center of the group $S_n$ is trivial for $n \geq 3$.*

***Proof*** This is due to Theorem 4.24. Two different permutations with the same cycle structure are conjugate, and therefore both are not in the center.          □

**Exercises**

1. Recalculate the order of the group of orientation-preserving isometries of the cube using the orbit-stabilizer theorem.
2. Prove that the centralizer of an element of an abelian group $G$ is all of $G$. The centralizer of a nontrivial rotation $d \in D_n$ is $D_n^+$ with one exception, which one?

3. Determine the conjugacy classes of the group $D_4$. You can use Theorem 4.24 for this, but be careful: Being conjugate in $D_4$ is not necessarily the same as being conjugate in $S_4$.
4. Assume the group $G$ acts on $X$, and $x, y \in X$ are in the same orbit, i.e., there exists a $g \in G$ with $y = g(x)$. Prove that the stabilizer of $y$ is the subgroup of the stabilizer of $x$ conjugated with $g$: $G(y) = gG(x)g^{-1}$ (for an example, see Theorem 3.45).
5. For $r \geq 0$, let $S_r^{n-1} = \{(x_1, \ldots, x_n) \in \mathbb{R}^n \mid x_1^2 + \cdots + x_n^2 = r^2\}$ be the sphere of dimension $n - 1$ in $\mathbb{R}^n$ with radius $r$ and the origin as center. Prove that the orbits of the action of $O_n$ on $\mathbb{R}^n$ are the spheres $S_r^{n-1}$.
6. Let $G$ be a group and $C(G)$ its center. Prove that

$$\bigcap_{g \in G} Z(g) = C(G).$$

## 4.5  Cayley Graphs

Cayley graphs are an important tool for representing finitely generated groups geometrically.

**Definition 4.46** A *graph* $\Gamma = (V, K)$ consists of an (at most countably infinite) set of vertices $V$ and a set of edges $K$, where each edge $k \in K$ runs between two vertices $v_1, v_2 \in V$.

In Fig. 4.5 we see a graph. Its vertex set is $\{a, b, c, d, e, f, g, h, i, j, k\}$. Edges can be described as pairs of vertices. The edge set of the graph is

$$\{\{a, b\}, \{b, c\}, \{c, d\}, \{d, f\}, \{f, j\}, \{j, k\}, \{k, c\}, \{d, e\}, \{f, g\}, \{g, h\}, \{f, i\}\}.$$

**Fig. 4.5**  A connected graph

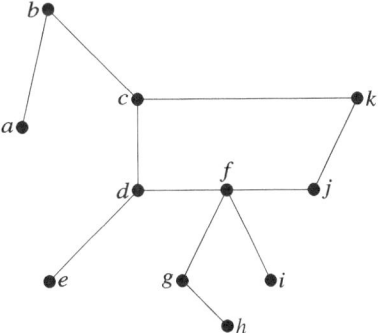

**Fig. 4.6**   A tree

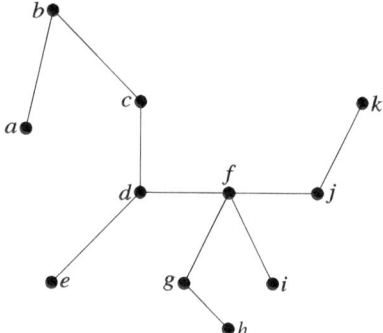

Let $\Gamma$ be a graph with vertex set $\{v_1, \ldots, v_n\}$ and edge set $\{e_1, \ldots, e_m\}$. A *path* in $\Gamma$ is a sequence of vertices and edges

$$\omega = (v_{i_1}, e_{i_1}, v_{i_2}, e_{i_2}, \ldots, v_{i_{k-1}}, e_{i_{k-1}}, v_{i_k}),$$

where vertices and edges alternate and the boundary vertices of $e_{i_j}$ are the vertices $v_{i_j}$ and $v_{i_{j+1}}$. The path $\omega$ is called *closed* if $v_{i_1} = v_{i_k}$. In the graph of Fig. 4.5 is

$$(b, \{b, c\}, c, \{c, k\}, k)$$

a non-closed path. The path

$$\alpha = (c, \{c, k\}, k, \{k, j\}, j, \{j, f\}, f, \{f, d\}, d, \{d, c\}, c)$$

is closed. A graph is called *connected* if there is a path between any two vertices. The graph in Fig. 4.5 is connected. A closed path is called *cycle* if no edge is traversed back and forth in direct succession. The path $\alpha$ is a cycle in the graph of Fig. 4.5.

**Definition 4.47**   A *tree* is a connected graph without cycles.

A tree is depicted in Fig. 4.6.

The *valency* (or *degree*) of a vertex is the number of edges that are incident to the vertex. The vertex $a$ has valency 1, and the vertex $f$ has valency 4 in the graph of Fig. 4.5.

A graph $\Gamma = (V, K)$ is called *oriented* if each edge $k \in K$ is provided with an orientation (a direction).

**Definition 4.48**   Let $G$ be a group with a generating system $\{g_1, \ldots, g_n\}$, so $G = \langle g_1, \ldots, g_n \rangle$. We assign to this group with generating system an oriented graph $\Gamma_G(g_1, \ldots, g_n)$: For each group element, we take a vertex, so the set of vertices is $G$. The vertices $h', h \in G$ are connected with an oriented edge from $h'$ to $h$ when $h'g_i = h$ for a $g_i \in \{g_1, \ldots, g_n\}$. The graph $\Gamma_G$ is called the *Cayley graph* of the group $G$ with respect to the generating system $\{g_1, \ldots, g_n\}$.

**Fig. 4.7** $\Gamma_{\mathbb{Z}}$ generated by 1

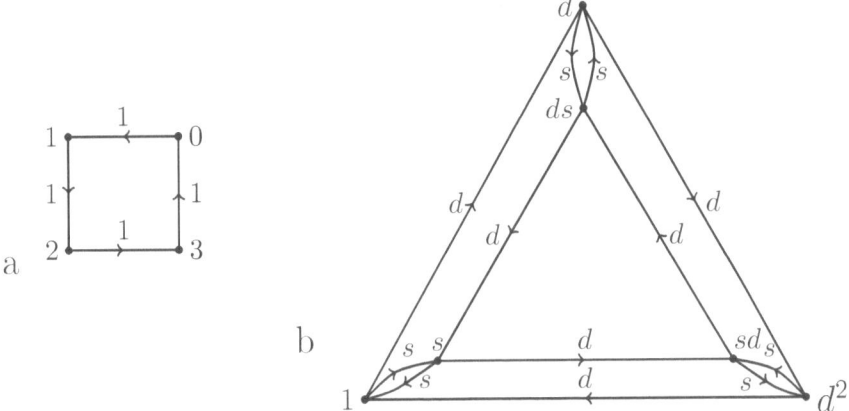

**Fig. 4.8** Cayley graphs of $(\mathbb{Z}_4, +_4)$ generated by 1 and $D_3$ generated by a reflection and a rotation

If it is clear which generating system is meant, we also write $\Gamma_G$ instead of $\Gamma_G(g_1, \ldots, g_n)$. We label the vertices of a Cayley graph with the corresponding group elements. Each edge is labeled with the generator that transforms the corresponding group elements into each other, i.e., the edge corresponding to $h' g_i = h$ is labeled with $g_i$. The degree of each vertex of the Cayley graph is $2n$ if the corresponding generating system consists of $n$ elements.

**Example 4.49** A part of the Cayley graph $\Gamma_{\mathbb{Z}}(1)$ for the group $\mathbb{Z}$ generated by 1 is shown in Fig. 4.7. Since the group $\mathbb{Z}$ is infinite, $\Gamma_{\mathbb{Z}}$ is also infinite. One must imagine this Cayley graph as extending infinitely to the right and left.

**Example 4.50** The Cayley graph for the group $(\mathbb{Z}_4, +_4)$ generated by 1 (see page 21) is shown in Fig. 4.8a.

**Example 4.51** The Cayley graph for the group $D_3$ generated by a reflection $s$ and a rotation $d$ by $120°$ is shown in Fig. 4.8b. Note that $sd = d^2 s$ and $ds = sd^2$.

In Cayley graphs which have reflections (or, more generally, involutions) as generators, we can save ourselves some drawing work by replacing pairs of forward and backward leading edges which are labeled with the same reflection by a non-oriented edge. So we can replace Fig. 4.8b by Fig. 4.9.

One can see left cosets very nicely in Cayley graphs. For example, we consider the subgroup $H < D_3$ with $H = \{id, s\}$ in Fig. 4.9. In the picture, it consists of the two vertices labeled 1 and $s$. The cosets $dH = \{d, ds\}$ and $d^2 H = \{d^2, sd = d^2 s\}$

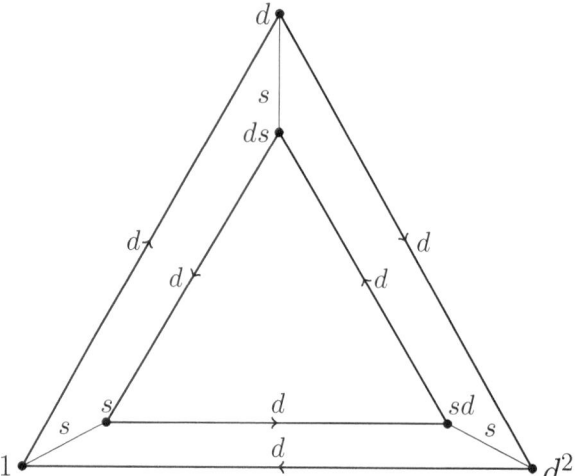

**Fig. 4.9** $D_3$ generated by a reflection and a rotation

can be seen geometrically as "rotated". Whether a subgroup is a normal subgroup can also be seen in the Cayley graph. Identify the cosets in the Cayley graph as single vertices and identify edges with the same starting and ending point. If this leads to a Cayley graph, the subgroup was a normal subgroup, and the Cayley graph is that of the factor group. If no Cayley graph is created, the subgroup is not a normal subgroup. The subgroup $H = \{id, s\}$ is not a normal subgroup (see Exercise 2. of Sect. 3.4). The graph created by drawing the cosets together to points is not a Cayley graph because edges labeled with $d$ lead back and forth between any two vertices.

However, the subgroup $N = \langle d \rangle$ is normal in $D_3$. It has two left cosets $N = \{1, d, d^2\}$ and $sN = \{s, ds, sd\}$. If you identify these to a point, you get two vertices that are connected with an edge labeled with $s$. This is the Cayley graph of the factor group isomorphic to $\mathbb{Z}_2$.

We also consider a Cayley graph of the group $S_4$. A generating system is $\{(1, 2), (2, 4, 3)\}$. We prove this with GAP:

The group generated by $(1, 2)$ and $(2, 4, 3)$ has 24 elements, and since there are only 24 permutations of the numbers $\{1, 2, 3, 4\}$, this must already be the whole group $S_4$.

```
gap> S4:=Group((1,2),(2,4,3));
Group([ (1,2), (2,4,3) ])
gap> Order(S4);
24
```

The Cayley graph $\Gamma_{S_4}((1, 2), (2, 4, 3))$ is depicted in Fig. 4.10. We imagine edges without orientation labeled with $(1, 2)$, the oriented edges with $(2, 4, 3)$. We omit the commas in the permutations for better readability.

Many Cayley graphs can be beautifully visualized with the software `Group Explorer` [Car19].

An important application of Cayley graphs is the following:

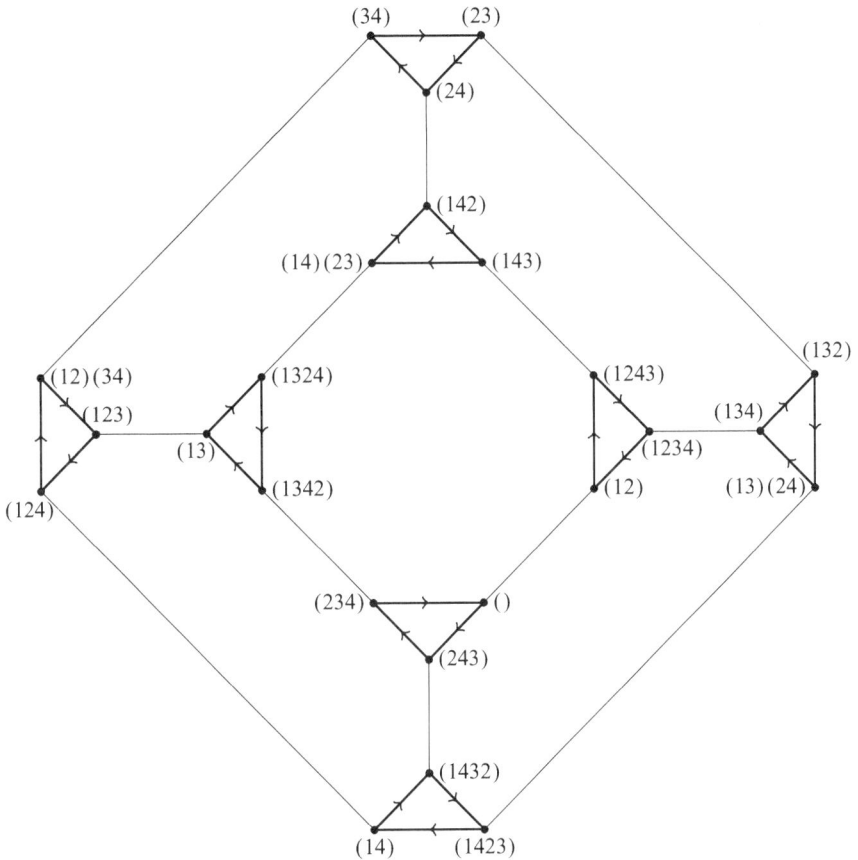

**Fig. 4.10** The Cayley graph of the group $S_4$ generated by $(1, 2)$, $(2, 4, 3)$

**Theorem 4.52** *Every finitely generated group operates on its Cayley graph.*

This operation corresponds to the one of Example 4.20. If $g \in G$ is a group element and $x \in \Gamma_G$ is a vertex of the Cayley graph (which is nothing else than a group element $x \in G$), then $g(x) = g \cdot x$, where $\cdot$ is the operation in the group $G$. This is compatible with the operation on the edges, because if $k \in \Gamma_G$ is an edge from the vertex $h'$ to $h$, which is labeled with $g_i$ (i.e. $h' \cdot g_i = h$), then $g(k) \in \Gamma_G$ is an edge from the vertex $g \cdot h'$ to $g \cdot h$, which is also labeled with $g_i$. From $h' \cdot g_i = h$ it follows that $g \cdot h' \cdot g_i = g \cdot h$.

Here are some examples concerning the Cayley graph of the group $D_3$ generated by $s$ and $d$ in Fig. 4.9. It is easy to see how $d$ acts on the Cayley graph:

$$d(1) = d, d(d) = d^2, d(d^2) = 1 \text{ and } d(s) = ds,$$

$$d(ds) = d^2 s = sd, d(sd) = dsd = s.$$

Each vertex is rotated clockwise by one vertex. How does $s$ operate? We have

$$s(1) = s, s(d) = sd, s(d^2) = sd^2 = ds \text{ and } s(s) = 1,$$

$$s(ds) = sds = d^2, s(sd) = d,$$

so 1 and $s$ swap their places, $d^2$ and $ds$ swap their places, and $d$ and $sd$ swap their places.

In the following, the operation of a group on its Cayley graph is interpreted as an operation on a geometric space. Initially, there is nothing geometric about a Cayley graph of a group. Without the ability to measure lengths, a graph is just a set (or several sets) and not a geometric space.

**Definition 4.53** A pair $(X, d)$ is called a *metric space* if $X$ is a set of points and $d$ assigns a distance to each pair of points. A *distance* is a function $d : X \times X \to \mathbb{R}_0^+$ which assigns a non-negative real number to each pair of points from $X$, namely their distance, and which fulfills the following conditions:

1. $d(P, Q) = 0$ if and only if $P = Q$,
2. $d(P, Q) = d(Q, P)$ for all $P, Q \in X$,
3. $d(P, R) \leq d(P, Q) + d(Q, R)$ for all $P, R, Q \in X$.

For example, the Euclidean plane is a metric space, where the distance function $d$ is defined as

$$d(p_1, p_2) = \sqrt{(x_1 - x_2)^2 + (y_1 - y_2)^2},$$

if the point $p_i$ is described by the coordinates $(x_i, y_i)$.

Condition 3 says that the shortest path from $P$ to $R$ cannot become shorter if we make a detour via $Q$. Such a function $d$ is also called a *metric* over $X$.

If $\Gamma_G(g_1, \ldots, g_n)$ is the Cayley graph of a finitely generated group $G$, we naturally obtain a metric $d_\Gamma$ on $\Gamma_G$ by assigning length 1 to each edge. This metric is called the *word metric*. The distance $d_\Gamma(w, v)$ in the word metric between two group elements $w, v \in G$ is the length of the shortest path from $w$ to $v$ in $\Gamma_G$, i.e., the length of the shortest word in the generators $g_1, \ldots, g_n$ which equals $wv^{-1}$ in $G$. Each word $w$ in the generators corresponds to a path in the Cayley graph from the vertex labeled 1 to the group element $w$. If $w$ and $v$ are two different words in the generators, but $w = v$ in the group, then we have two different paths to the same vertex in the Cayley graph. The Cayley graph $\Gamma_G$ of a group $G$ is thus a metric space $(\Gamma_G, d_\Gamma)$.

The operation of Theorem 4.52 is thus an operation on a metric space. If we let any element $g$ of the group operate on $\Gamma_G$, this is a bijective mapping of the Cayley graph onto itself. Each point and each edge of $\Gamma_G$ has exactly one point or edge in the preimage, because if $g(x) = g(x')$, then $gx = gx'$ and thus $x = x'$ follows. The surjectivity is just as clear. But this bijective mapping is even distance-preserving: If $w$ is a *line segment*, i.e., a shortest connection, of the points $v$ and $vw$ in $\Gamma_G$, then

$g(w)$ is a line segment between the points $g(v)$ and $g(vw)$ of the same length. So we have proven:

**Theorem 4.54** *Let $G$ be a finitely generated group with a finite generating system $X$. Let $d_\Gamma$ be the word metric on $\Gamma_G(X)$. Then $G$ acts by isometries on the metric space $(\Gamma_G(X), d_\Gamma)$.*

In this light, it is not surprising that the Cayley graph of the group $D_3$ of Fig. 4.9 looks so similar to an equilateral triangle. Each element of the group $D_3$, i.e. each isometry of the triangle, acts on the Cayley graph. For example, the operation with the group element $d$ on the Cayley graph in Fig. 4.9 corresponds to a clockwise rotation of the Cayley graph by $120°$. The vertex $d^2$ changes into the vertex 1, and $s$ changes to $ds$. One must handle this interpretation with some caution, as the Cayley graph should not be imagined as embedded in the plane, which would allow the interpretation as a rotation.

There is a very important theorem in modern group theory due to Švarc and Milnor [Mil68], which roughly states: If $G$ is a finitely generated group that operates on a figure $F$ by isometries (in a particularly beautiful way), then $F$ is "almost the same" as the Cayley graph of $G$. This theorem was one of the turning points of modern geometric group theory and led to a view of groups as geometric objects and thus to developments such as hyperbolic and automatic groups. We explain this theorem in more detail in Sect. 10.2.

To make clear how the figure and the Cayley graph are related, we consider an example in Sect. 4.6.

**Exercises**

1. Draw the Cayley graph for the group $D_5$ and then generally for the group $D_n$, each generated by a reflection and a rotation. Make the operation of $D_5$ on its Cayley graph geometrically clear. To do this, let a rotation and a reflection operate on $\Gamma_{D_5}$, and consider the images of each vertex and edge under this operation.
2. Draw the Cayley graph for the symmetry group of the rhombus generated by the two reflections (see Definition 2.20).
3. Prove that the operation of a group on the vertices and edges of its Cayley graph of Theorem 4.52 is transitive.
4. Prove that Cayley graphs are connected.
5. Let $\tau_1$ and $\tau_2$ be two translations of perpendicular vectors of length 1 in the plane. Let $\mathbb{Z} \times \mathbb{Z}$ be the group generated by $\tau_1$ and $\tau_2$ (the notation is explained in Sect. 6.1). Draw (a part) of the associated Cayley graph.
6. Prove that if $G$ is a finite group, then in every Cayley graph $\Gamma_G$ for all $g, h \in G$ the inequality $d_\Gamma(g, h) \leq |G|$ holds.
7. Draw the Cayley graph for $\mathbb{Z}$ generated by $\{2, 3\}$.

## 4.6   A Decomposition of the Plane

If one has a figure in the plane (or a solid in space) and wants to draw the Cayley graph of the associated symmetry group, there is a trick for this. One chooses a point in the figure with a trivial stabilizer and draws the entire orbit. Corollary 4.37 tells us that this will already give us the vertices of the Cayley graph. The edges may also be easily found. We will consider an example in detail.

Consider the *decomposition* $Z$ of the plane into equilateral triangles of Fig. 4.11. Let $G = G_{(3,6)}$ be the symmetry group of this decomposition. $G$ operates on the set of all equilateral triangles of the decomposition. The notation $G_{(3,6)}$ comes from the fact that we have divided the plane into regular triangles, arranged so that 6 triangles meet at each vertex.

**Definition 4.55**  Let $X$ be a $G$-set. If for each $x \in X$ and each nontrivial group element $g \in G$, $g \neq id$, the relation $g(x) \neq x$ holds, then the operation is called *free*.

Obviously, a group operation is free if the stabilizer of each $x \in X$ is trivial. The operation of a group on its Cayley graph of Theorem 4.52 is free, because each nontrivial group element $g$ transports a group element $x$ to another group element $gx$ and the attached edges with it.

In the example of the set of triangles of Fig. 4.11, the operation is not free, because the reflection over a mirror axis, which halves a triangle, maps this triangle onto itself.

We draw all mirror axes in Fig. 4.11 and obtain Fig. 4.12. We now have a decomposition $Z'$ of the plane into right-angled triangles. $G$ operates freely on $Z'$. Every reflection or rotation which maps the original decomposition $Z$ onto itself (except for the identity) maps a right-angled triangle from $Z'$ onto another right-angled triangle.

Our goal is to find the Cayley graph of $G$ with respect to a generating system of $G$ in $Z'$. To do this, we first define a *fundamental domain* for $G$, which is a region of the plane whose images under the elements of $G$ are different and fill the entire plane in such a way that these images overlap at most at edges and vertices.

**Fig. 4.11**   A decomposition $Z$ of the plane

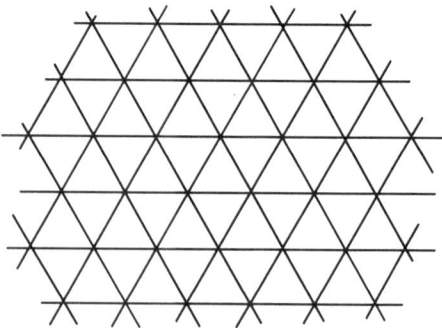

**Fig. 4.12**  A decomposition
$Z'$ of the plane into
right-angled triangles

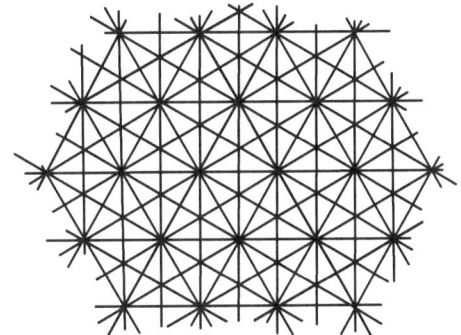

**Fig. 4.13**  The fundamental
region $f$ in the
decomposition $Z'$

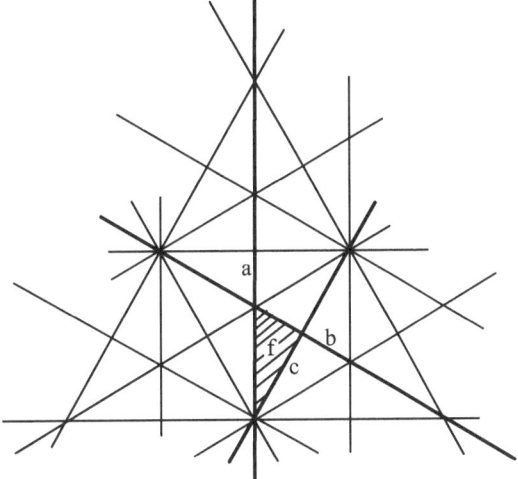

An arbitrary, but from now on fixed triangle $f \in Z'$ including its boundary edges
and vertices serves as a fundamental domain: $f$ is part of a triangle $d \in Z$, and
the stabilizer of $d$ in $G$ contains exactly the group elements that map $f$ onto the
other (right-angled) triangles in $d$. Reflections and translations of $G$ then map $d$
onto the other equilateral triangles. With a translation, every second triangle of $Z$ is
reachable, namely those that have their tip pointing downwards, like the triangle $d$
(see Fig. 4.13). All other triangles of $Z$ can be reached with an additional upstream
reflection over a boundary edge of $d$.

The images of $f$ under $G$ thus fill the entire plane. Corollary 1.8 proves that two
elements of $G$ that map $f$ onto the same right-angled triangle $f'$ must be equal.
Let $a, b, c$ be the three lines that bound $f$, as shown in Fig. 4.13. Let $s_a \in G$ be
the reflection over the line $a$, $s_b \in G$ in $b$ and $s_c \in G$ in $c$. The following theorem
can be literally *seen*, if we set up three mirrors perpendicular to each other, so that
they form a triangle with interior angles of 30, 60 and 90°. These mirrors represent
reflections $s_a$, $s_b$ and $s_c$. If one then looks from above into the mirrors, one sees

**Fig. 4.14** A part of the decomposition $Z'$

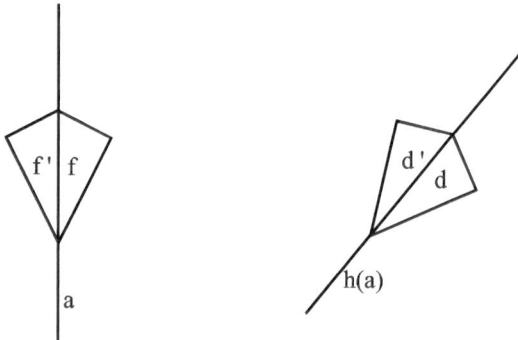

the decomposition $Z'$. With these three mirrors, by repeated reflection, every mirror axis of $Z'$ is obtained, i.e.:

**Theorem 4.56** $G$ *is generated by* $s_a, s_b, s_c$.

**Proof** Let $H < G$ be the subgroup generated by $s_a, s_b, s_c$. We want to prove $H = G$ and assume that $H \neq G$. Because $f$ is a fundamental region, $Z'$ is the union of all $g(f)$ for all $g \in G$.

From $H \neq G$ it follows that $\{h(f) \mid h \in H\}$ does not completely cover $Z'$. Therefore, there must be two triangles with a common edge, where one triangle $d \in Z'$ is reached by an isometry from $H$ and the second triangle $d' \in Z'$ is not reachable by an isometry from $H$. So there must be 2 triangles $d, d' \in Z'$ with a common mirror axis $s$ on their boundary, so that there is an isometry $h \in H$ for one triangle $d$ with $h(f) = d$, but not for the other triangle $d'$. But this isometry $h$ also maps the boundary of $f$ to the boundary of $d$, so $s$ is the image of $a, b$ or $c$ under $h$. We assume $h(a) = s$ (the other two cases are analogous). See Fig. 4.14.

The mapping $h s_a h^{-1}$ maps the triangle $d$ to $d'$:

$$h s_a h^{-1}(d) = h s_a(f) = h(f') = d',$$

where $f'$ is the corresponding neighboring triangle of $f$. Note that we always perform isometries from right to left.

But this is a contradiction, because if $h, s_a \in H$ then also $h s_a \in H$ and because $h s_a(f) = d'$ the element $d'$ is reachable with elements from $H$.  □

The exact same proof works in general:

**Theorem 4.57** *Let* $Z \subset \mathbb{R}^n$ *be a decomposition with symmetry group G. Consider the decomposition* $Z'$ *which is induced by the hyperplanes belonging to all reflections of G. If G operates freely on* $Z'$, *then G is generated by the reflections on the boundary hyperplanes of a tile.*

As a simple example, we consider the regular $n$-gon embedded in the plane. The $n$ mirror axes of the $n$-gon divide the plane into congruent parts. Any one of these parts serves as a fundamental region. The two reflections on the boundary of this fundamental region generate the group $D_n$.

Back to our decomposition of the plane: We now *dualize* the decomposition $Z'$, i.e., we draw a vertex in the center of each triangle. Two vertices are connected with an edge if the corresponding triangles share an edge.

Then we assign a group element to each vertex and each edge. The vertex labeled with the identity element 1 is assigned to the fundamental region $f$. An edge of this dual graph receives the label $s_a$ or $s_b$ or $s_c$, depending on whether the original edge $s$ of the decomposition $Z'$ is the image of $a$, $b$ or $c$ under the operation of $G$. The resulting graph becomes the Cayley graph $\Gamma_G(s_a, s_b, s_c)$. The designation of the other vertices of the Cayley graph is automatically determined by the edge labels. The edges are not oriented because all three generators are reflections and, as in the example of the group $D_3$, we replace each pair of oriented edges with the same reflection by a non-oriented edge.

We thus obtain Fig. 4.15. For better clarity, the edges of the Cayley graph are drawn dashed, and the labels of vertices have been omitted. We will include the latter in Fig. 4.16 for some vertices.

If one looks at the labels of the vertices, one finds that for a given group element $g$, the fundamental domain $f$ is mapped by $g$ onto the triangle with vertex label $g$. The reader is recommended to check the vertex labels, i.e. to mentally map the fundamental domain $f$ with the figure standing at a vertex onto the corresponding triangle. We check this in general: $f$ is mapped by $s_a, s_b, s_c$ onto the respective neighboring triangles. We inductively assume that, for all triangles with distance at most $n$ from $f$: If a triangle $d$ has the vertex label $g$, then $f$ is mapped onto $d$ by the operation $g$. The distance between two vertices of the Cayley graph is the number of edges on the shortest path between the two vertices.

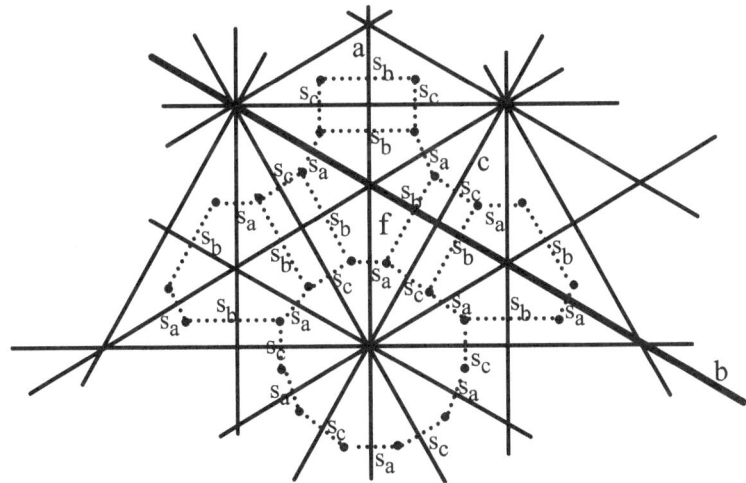

**Fig. 4.15** Cayley graph of $G$ in the decomposition $Z'$

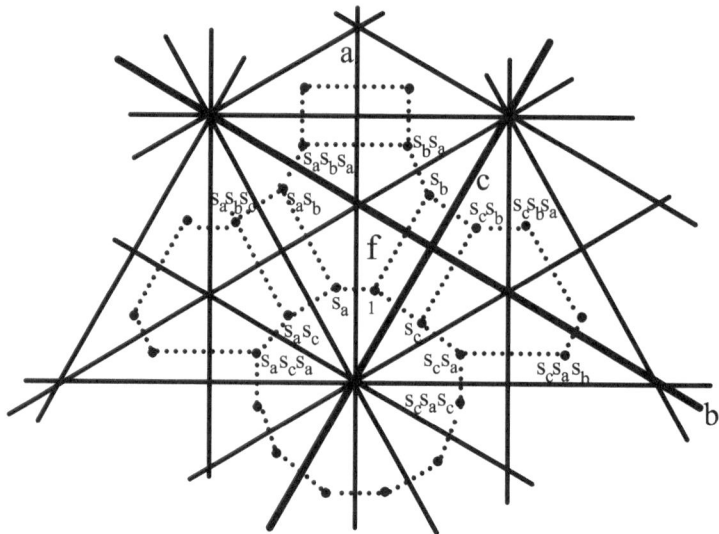

**Fig. 4.16**  Cayley graph of $G$ in the decomposition $Z'$ with labeling of the vertices

Let $P$ be a vertex at distance $n$ from the vertex 1 with the label $g_n$. Inductively we know that $g_n(1) = P$. Let $P'$ be a neighboring point of $P$ with the label $g_n h$. So there is an edge with the label $h$ between $P$ and $P'$. We have to show $g_n h(1) = P'$. But this is true because $h$ maps 1 onto the vertex with the label $h$, which in turn is mapped onto $P'$ by the operation $g_n$.

Now it is also clear why the Cayley graph results from dualizing. We get for each group element a vertex, and the operation on the edges corresponds to that of Theorem 4.52.

**Exercises**

1. Complete the vertex labels in Fig. 4.16.
2. Consider the decomposition $Q$ of the plane into congruent squares of Fig. 1.4 with symmetry group $G_{(4,4)}$, and follow Sect. 4.6 for $Q$. So find a fundamental domain $f$ for $G_{(4,4)}$, by further subdividing the plane into a decomposition $Q'$, on which $G_{(4,4)}$ operates freely. Prove Theorem 4.31 anew (see page 74). So show that $G_{(4,4)}$ is generated by the reflections on the boundary lines of $f$, and draw the Cayley graph for this generating system of the group $G_{(4,4)}$.
3. Consider again a decomposition $Q$ of the plane into congruent squares. Let $G$ be the group generated by the boundary lines of a single square. Show $G \triangleleft G_{(4,4)}$. Determine $G_{(4,4)}/G$.
4. Prove that the decomposition of the plane into regular hexagons has the same symmetry group as the decomposition into regular triangles, by showing that the respective fundamental regions are the same.

5. Find the Cayley graph for the group $D_3$ generated by two reflections. To do this, find the (unbounded) fundamental region in Fig. 1.1, and draw the Cayley graph in (a copy of) this figure.

6. Prove that the symmetry group of the tetrahedron has a subgroup $H < S_4$ generated by $(1, 2, 4, 3)$, which operates on the sides of the tetrahedron with the side $d$ as the fundamental region, where $d$ is the side with the vertices 1,2,3. We obtain $S_4 = S_4(d) \cdot H$. Neither $S_4(d)$ nor $H$ is normal in $S_4$.

7. Follow Theorem 4.57 with mirrors for further examples. For this, for example, place two mirrors adjacent to each other at an angle of $360/2n$ degrees. The enclosed angle range serves as the fundamental region for the group $D_n$. You can see the $n$ mirror axes of the regular $n$-gon.

# Chapter 5
# Group Presentations

Groups can be represented by generators and relations, which are products of generators and their inverses that yield the identity element. This representation is called a presentation of a group and describes it completely. Conversely, however, it is generally not possible to decide from two given presentations whether they describe the same group or not. This leads us to the decision problems, first formulated by Max Dehn, which can be solved in many cases in a very geometric way as described in Chap. 10.

## 5.1 Group Presentations

Let the group $D_3$ be given by

$$D_3 = \{id, d_{120}, d_{240}, s_a, s_b, s_c\}.$$

In Exercise 5. of Sect. 2.1 we saw that the group $D_3$ can be generated by any reflection $s = s_a$ and a rotation $d = d_{120}$ by $120°$, i.e. $D_3 = \langle d, s \rangle$. Every element of $D_3$ can thus be written as a product of $d$ and $s$ and their inverses. For example, for the rotation by $240°$ $d_{240} = d^2$, as can be seen in Fig. 1.1. We have

$$D_3 = \{id, s, d, d^2, sd, sd^2\}.$$

(We already clarified this in Sect. 3.2.) A *word* is any expression in the generators and their inverses, in our case an expression in $s, d, s^{-1}, d^{-1}$. Examples of words are $s^2 d^{-3} sd$ or $sdsd$. There are certain words that are trivial in the given group. Because $d$ is a rotation by $120°$, $d^3 = id$, and $s$ is a reflection, so $s^2 = id$. Such words are called relations.

© The Author(s), under exclusive license to Springer-Verlag GmbH, DE,
part of Springer Nature 2024
S. Rosebrock, *Visual Group Theory*, Springer Undergraduate Mathematics Series,
https://doi.org/10.1007/978-3-662-69365-0_5

**Definition 5.1** Let $G = \langle g_1, \ldots, g_n \rangle$ and $w$ be a word in the generators $g_1, \ldots, g_n$ and their inverses. If $w = 1$ in $G$, then $w$ is called a *relation* or *relator* in $G$.

If the words $w$, $v$ are exactly the same words, we write $w \equiv v$. If they are only equal in the group under consideration, we write $w = v$ like $s^4 = 1$ for a reflection $s \in D_3$. The words $s^2$ and $d^3$ are thus relations in $D_3 = \langle s, d \rangle$. Of course, $s^4$ is also a relation. This can be derived from $s^2$.

$s^2 d^{-2} dss^{-1} ds^{-2}$ is a relation that can be reduced to 1 by shortening pairs of $ss^{-1}$ and $dd^{-1}$ (we often write 1 for *id* in this context). Such relations are called *freely reducible* to the identity element 1. We will not list these in lists of relations. A word is called *reduced* when such shortenings of pairs of generators and their inverses are not possible.

Relations can be very easily recognized in the Cayley graph. Let $w$ be a word in the generators of a group $G = \langle g_1, \ldots, g_n \rangle$. We can thus write $w$ as $w \equiv a_1 a_2 \ldots a_n$, where each $a_i$ is a generator $g_j$ or inverse of a generator $g_j^{-1}$. For $w$ we find a path in the Cayley graph $\Gamma_G(g_1, \ldots, g_n)$, by starting at 1 and first going along the edge with the label $a_1$ with its orientation. We then land at the vertex $a_1$. Then we go along the edge with the label $a_2$, which takes us to the vertex $a_1 a_2$. If $a_i$ is an inverse generator, we go against the direction of the arrow and otherwise we go with the direction of the arrow. For each word $w$ there is therefore a path $\gamma_w \subset \Gamma_G$. We can recognize relations in the Cayley graph as follows:

**Theorem 5.2** *Let* $G = \langle g_1, \ldots, g_n \rangle$ *and* $w \equiv a_1 a_2 \ldots a_m$ *be a word in the generators and their inverses. Then* $w$ *is a relation in* $G$ *if and only if* $\gamma_w \subset \Gamma_G(g_1, \ldots, g_n)$ *is closed.*

**Proof** Let $w$ be a relation. So $w = 1$ in $G$. By the definition of the Cayley graph, we arrive at the vertex with label $a_1$ if we walk from the vertex with label 1 along the edge with the label $a_1$. From the vertex $a_1$ we walk along the edge $a_2$, which takes us to the vertex $a_1 a_2$. If we have completely walked $w$, we are at the vertex $a_1 a_2 \ldots a_m$. Since $w$ is a relation, this word is equal to 1, which is the label of our starting point. So the path $\gamma_w$ is closed.

Conversely, let $\gamma_w$ be a path that starts and ends at the vertex $v \in G$. This path reads a word $a_1 a_2 \ldots a_m$ along its edges. Since the path ends again at $v$, it follows that $v = v a_1 a_2 \ldots a_m$ and thus $a_1 a_2 \ldots a_m = 1$.                                    □

Back to our example: We have $s_c = s_a d_{120} = sd$, and $s_c$ has order 2. So $sdsd$ is a relation, which we recognize as a closed path, as well as the relation $d^3$, in Fig. 4.9 on page 84. The relation $s^2$ is implicitly in the figure, because we have replaced oriented pairs of edges by unoriented edges.

We say a relation $w = 1$ can be *derived* from a set of relations $\{w_1, \ldots, w_m\}$, or is *derivable* from these, if the equation $w = 1$ can be inferred by applying the relations $\{w_1, \ldots, w_m\}$ and free reductions. The three relations $s^2$, $d^3$ and $sdsd$ form a set of defining relations, meaning that every other relation is derivable from these.

The relation $s^4 = 1$ can be derived from $s^2$ by applying the relation $s^2$ twice in a row. The relation $w \equiv d^2 s d^2 s$ can also be derived from the defining relations:

The defining relation $sdsd = 1$ can be transformed into $sdsd^3 = d^2$ by multiplying from the right by $d^2$. From the defining relation $d^3$ it follows that $sds = d^2$ (we have already established this in Sect. 4.4). If we replace the two $d^2$ in $w$ with $sds$ it follows that $w = sdsssdss$. From the defining relation $s^2$ we get $w = sdsd$, which is itself a defining relation. So we have transformed the word $w = d^2sd^2s$ into the identity only by applying defining relations.

**Definition 5.3** Let $G = \langle g_1, \ldots, g_n \rangle$, and $w_1, \ldots, w_m$ be relations in $G$. If every other relation can be derived from $\{w_1, \ldots, w_m\}$, then the set $\{w_1, \ldots, w_m\}$ is called a set of *defining relations* of $G$ with respect to the generators $g_1, \ldots, g_n$.

We still have to prove that $\{s^2, d^3, sdsd\}$ is a set of defining relations for $D_3 = \langle s, d \rangle$. Before we do that, some more theory:

**Definition 5.4** Let $G = \langle g_1, \ldots, g_n \rangle$, and $w_1, \ldots, w_m$ be defining relations in the generators $g_1, \ldots, g_n$. Then $\langle g_1, \ldots, g_n \mid w_1, \ldots, w_m \rangle$ is called a *presentation* of the group $G$ and we write $G = \langle g_1, \ldots, g_n \mid w_1, \ldots, w_m \rangle$.

In principle, a group can require infinitely many generators and relations. Those groups are called *infinitely generated*. We have already seen in Theorem 2.23 that the group $(\mathbb{R}, +)$ is infinitely generated. Similarly, the symmetry group of the circle is infinitely generated. This is because there are arbitrarily small real numbers and rotations by arbitrarily small angles. These two groups require infinitely many relations in any presentation and are therefore called *infinitely presented*. There are also groups that manage with finitely many generators but require infinitely many relations.

Now the promised proof in general for all dihedral groups:

**Theorem 5.5** $D_n = \langle s, d \mid s^2, d^n, sdsd \rangle$.

**Proof** As we have seen above, $s^2$, $d^n$ and $sdsd$ are relations in $D_n$. We still have to show that if $w$ is any relation in $D_n$, it can be transformed into the group identity using the defining relations. We look again at the representation of the dihedral group $D_n$ of Sect. 3.2. There we found that the elements of $D_n$ can be written as

$$D_n = \{id, d, d^2, \ldots, d^{n-1}, s, sd, sd^2, \ldots, sd^{n-1}\}. \tag{5.1}$$

This is a *normal form* for the elements of the group $D_n$, i.e., an arbitrary word $w$ in the generators $s, d$ can be transformed into one of the words from (5.1) using only the above three defining relations. Then relations are transformed into the group identity under this transformation.

Each word $w$ in $s, d$ has the form $w \equiv s^{\epsilon_1}d^{\delta_1}s^{\epsilon_2}d^{\delta_2}\ldots s^{\epsilon_k}d^{\delta_k}$. Because of the defining relation $s^2$, we can assume that for $i \neq 1$ each $\epsilon_i = 1$. Otherwise, we insert the defining relation $s^2$ often enough until we either get $s^1$ or $s^0$. $s^0 = 1$ can be omitted, and in that case our word becomes shorter. Similarly, we can assume that $\delta_i \in \{1, 2, \ldots, n-1\}$. $\epsilon = \epsilon_1$ or $\delta_k$ can also be 0. Our word is now $w = s^{\epsilon}d^{\delta_1}sd^{\delta_2}\ldots sd^{\delta_k}$, where $\epsilon \in \{0, 1\}$.

If we insert the defining relation $sdsd$ after the second $s$ into itself (we are allowed to do this because the defining relation $sdsd$ is a relation in $D_n$), we get $sds\,sdsd\,d$, or, after shortening $s^2$, the word $sd^2sd^2$. By inserting another $sdsd$ after the second $s$ we get $sd^3sd^3$ etc. So we can derive any word $sd^i sd^i$ from the defining relations.

The relation $sd^i sd^i = 1$ can also be written as $sd^i = d^{-i}s$. In our word

$$w = s^\epsilon d^{\delta_1} s d^{\delta_2} \ldots s d^{\delta_k}$$

we can therefore replace $sd^{\delta_2}$ with $d^{-\delta_2}s$ and get

$$w = s^\epsilon d^{\delta_1 - \delta_2 + \delta_3} s d^{\delta_4} \ldots s d^{\delta_k}.$$

We proceed analogously for the other subwords $sd^{\delta_i}$ until we get $w = s^\epsilon d^\tau$, where $\tau$ is the corresponding alternating sum of the $\delta_i$. After normalizing the $d$-power by using the defining relation $d^n$, $w$ has the form of a word from (5.1).     $\square$

With some practice, we can recognize the statement of Theorem 5.5 more easily by looking at the Cayley graph. Each relation is a closed path, and an arbitrary closed path in Fig. 4.9 must read along its edges $s^2$, $d^3$ or $sdsd$, or it must be composed of these elementary paths. For example, the path $sddsd^{-1}$ is closed and composed from $d^3$ and $(dsds)^{-1}$. In fact, in a Cayley graph which lies entirely in the plane, as in our example, one can particularly easily recognize a set of defining relations. The graph divides the plane into regions, and reading the cycle of edges around each region will give a relation of the set of defining relations. This can be easily checked in this example.

In GAP you first define the generators and then generate with

$$Generator/[Relations]$$

a presentation.

```
gap> F := FreeGroup("s","d");;
gap> AssignGeneratorVariables(F);
#I  Assigned the global variables [ s, d ]
gap> D7:= F/[s^2, d^7, (s*d)^2];;
gap> Elements(D7);
[ <identity ...>, s, s*d, d, s*d*s, d*s, s*d^2, d^2,
  s*d^2*s, d^2*s, s*d^3, d^3, s*d^3*s, d^3*s ]
```

Note that the elements are not necessarily in the normal form (5.1). (Exercise: Bring the elements into normal form.)

Presentations of groups are not unique:

**Theorem 5.6**  $D_n = \langle s_1, s_2 \mid s_1^2, s_2^2, (s_1 s_2)^n \rangle$.

**_Proof_** We consider the reflection $s$ and the rotation $d$ of Theorem 5.5 and set $s_1 = s$ and $s_2 = sd$ ($s_1$ and $s_2$ are thus two "adjacent" reflections in the regular $n$-gon).

Then from $s_1^2 = 1$ the relation $s^2 = 1$ follows, and from $s_2^2 = 1$ we get the relation $sdsd = 1$. From $(s_1 s_2)^n$ follows $d^n = 1$ (after crossing out $s^2$). So we can derive the defining relations of Theorem 5.5 from the relations $s_1^2, s_2^2, (s_1 s_2)^n$, and thus $s_1^2, s_2^2, (s_1 s_2)^n$ themselves are a set of defining relations.                                  □

**Example 5.7** The symmetry group of the rhombus of Fig. 1.5 (see also Example 2.8) has the presentation $\langle a, b \mid a^2, b^2, ab = ba \rangle$.

We check this with GAP:

```
gap> F := FreeGroup("a","b");;
gap> AssignGeneratorVariables(F);
#I  Assigned the global variables [ a, b ]
gap> R:=F/[a^2, b^2, a*b*a^-1*b^-1];
<fp group on the generators [ a, b ]>
gap> Elements(R);
[ <identity ...>, a, b, a*b ]
```

Indeed, the rhombus group has 4 elements, as we have established in Example 2.8, two reflections $s_a$ and $s_b$ and the product of these two reflections, which define a rotation by $180°$. The relation $ab = ba$ is encoded in GAP by $a*b*a^-1*b^-1$. This term is obtained by multiplying each side of $ab = ba$ from the right by $(ba)^{-1} = a^{-1} b^{-1}$.

We once again consider the group $G = G_{(3,6)}$ of Sect. 4.6, which operates on the decomposition of the plane into regular triangles. We are looking for a presentation of $G$. According to Theorem 5.2 we only need an overview of the closed paths in the Cayley graph $\Gamma_G(s_a, s_b, s_c)$, where $s_a, s_b, s_c$ are reflections over the lines $a, b, c$ of Fig. 4.13.

From Fig. 4.15 on page 91 we can immediately see that all closed paths can be composed from those of the form $(s_c s_b)^2$, $(s_a s_b)^3$ and $(s_a s_c)^6$. We therefore conclude

**Theorem 5.8** $G_{(3,6)} = \langle s_a, s_b, s_c \mid s_a^2, s_b^2, s_c^2, (s_c s_b)^2, (s_a, s_b)^3, (s_a s_c)^6 \rangle$.

**Exercises**

1. Prove the assertion of Example 5.7 without using GAP, either by calculation or by arguing using the Cayley graph.
2. Find at least two more presentations of the symmetry group of the rhombus.
3. Prove that the symmetry group of the frieze of Fig. 1.3 on page 3 has the presentation $P = \langle s, \tau \mid s^2 = \tau, \tau s = s\tau \rangle$, where $s$ is a glide reflection and $\tau$ is a translation (by twice the length of $\vec{v}$ of Fig. 1.3). Prove that $P$ is a presentation of the group $\mathbb{Z}$.
4. Show, using the example of the group $D_5$, how to obtain the Cayley graph from a presentation of a finite group. A presentation for the group $D_5$ is given in Theorem 5.5.
5. Find a presentation of the finite cyclic group $\mathbb{Z}_n$.

6. The *quaternion group* is a group of order 8 with the elements $Q =$ $\{\pm 1, \pm k, \pm i, \pm j\}$ and the relations

$$k^2 = i^2 = j^2 = -1, ij = k, jk = i, ki = j, ji = -k, kj = -i, ik = -j.$$

Show that

(a) $Q$ has order 8 (possibly with GAP),
(b) $Q$ has only one element of order 2 and that the center of $Q$ is generated by this element,
(c) $D_4 \not\cong Q$,
(d) every subgroup of $Q$ is normal,
(e) $Q = \langle x, y \mid x^4, x^2 y^{-2}, xyxy^{-1} \rangle$.

## 5.2  Free Groups

Free groups were introduced by WALTER VAN DYCK in 1882.

**Definition 5.9** A group that has a presentation without defining relations is called *free*. The set of generators of such a presentation is called a *basis* of the free group, and its size is the *rank* of the free group.

This is where the GAP command `FreeGroup` comes from. Alternatively, we can also define: A group is free if it has a presentation in which every relation is freely reducible to 1. A typical such word is for example $w \equiv a^3 b^2 cc^{-1} b^{-1} a^{-2} a^2 b^{-2} ba^{-3}$.

We have $\mathbb{Z} = \langle t \mid \rangle$. The integers thus form a free group of rank 1.

If one allows infinitely many generators, there are also free groups of infinite rank.

If two free groups have the same rank, they are isomorphic. The isomorphism is obtained by extending any bijective mapping between the two bases to an isomorphism of the groups. The group elements have only received new names, otherwise the groups are exactly the same.

Conversely, one can show that two isomorphic free groups have the same rank. The proof is more difficult, and we omit it.

According to Theorem 5.2, one recognizes Cayley graphs of free groups (with respect to a basis) by the fact that they have no cycles. The freely reducible words, such as the word $w$ above, correspond to closed paths in a tree. If the pair $aa^{-1}$ appears in a word, this means for the path in the Cayley graph that an edge labeled with $a$ is traversed back and forth directly in succession. Since Cayley graphs are always connected (see Exercise 4. of Sect. 4.5), this implies the following:

**Theorem 5.10** *The Cayley graph of a finitely generated free group with respect to a basis without relations is a tree.*

A free group thus operates freely on a tree. Even the converse is true (see [Arm88]): If a group operates freely on a tree, this group is free. With this we can easily prove the following theorem of *Nielsen–Schreier*:

**Theorem 5.11** *Subgroups of free groups are free.*

**Proof** Let $H < F$, and $F$ be free. According to Theorem 5.10, $F$ acts freely on a tree. Each of its subgroups operate freely on the same tree. By the converse of Theorem 5.10, this means that $H$ is free. □

Free groups have a *universal mapping property*. This means that if one maps the basis of a free group arbitrarily into another group, then one can also map all other group elements in such a way that one obtains a homomorphism. The following theorem clarifies this statement:

**Theorem 5.12** *Let $F$ be a group, and let $X \subset F$ generate $F$. Then the following statements are equivalent:*

1. *$F$ is free with basis $X$ (we say $F$ is free over $X$).*
2. *Every function $f : X \to G$ into an arbitrary group $G$ can be uniquely extended to a homomorphism $\phi : F \to G$.*

**Proof** 1⇒2:

Let $a \in F$ be arbitrary. We write $a$ as a product of the generators and their inverses:

$$a = x_{i_1}^{\epsilon_1} x_{i_2}^{\epsilon_2} \ldots x_{i_m}^{\epsilon_m}$$

with $x_{i_k} \in X$ and $\epsilon_k = \pm 1$. If

$$a = x_{j_1}^{\delta_1} x_{j_2}^{\delta_2} \ldots x_{j_p}^{\delta_p}$$

is another such representation, then

$$x_{i_1}^{\epsilon_1} x_{i_2}^{\epsilon_2} \ldots x_{i_m}^{\epsilon_m} (x_{j_1}^{\delta_1} x_{j_2}^{\delta_2} \ldots x_{j_p}^{\delta_p})^{-1}$$

is freely reducible to 1, because $F$ is free. It follows that

$$f(x_{i_1}^{\epsilon_1}) f(x_{i_2}^{\epsilon_2}) \ldots f(x_{i_m}^{\epsilon_m}) f(x_{j_p}^{-\delta_p}) \ldots f(x_{j_1}^{-\delta_1})$$

is freely reducible to 1. We can indeed perform the same reductions in the image as in the preimage. So

$$f(x_{i_1}^{\epsilon_1}) f(x_{i_2}^{\epsilon_2}) \ldots f(x_{i_m}^{\epsilon_m})$$

depends only on $a$. We define $\phi : F \to G$ by

$$\phi(a) = f(x_{i_1}^{\epsilon_1}) f(x_{i_2}^{\epsilon_2}) \ldots f(x_{i_m}^{\epsilon_m})$$

for each element $a \in F$. It is easy to see that $\phi$ is a homomorphism. This homomorphism is unique, since every homomorphism must have as an image of $a$

$$f(x_{i_1}^{\epsilon_1})f(x_{i_2}^{\epsilon_2}) \dots f(x_{i_m}^{\epsilon_m}).$$

**2⇒1:**

Let $f: X \to G$ be a function into a group $G$, and

$$a = x_{i_1}^{\epsilon_1} x_{i_2}^{\epsilon_2} \dots x_{i_m}^{\epsilon_m} = 1$$

be any element in $F$, where the $x_i$ are in $X$. Since $f$ can be extended to a homomorphism $\phi: F \to G$, it follows that $\phi(a) = \phi(x_{i_1}^{\epsilon_1} x_{i_2}^{\epsilon_2} \dots x_{i_m}^{\epsilon_m}) = \phi(x_{i_1}^{\epsilon_1})\phi(x_{i_2}^{\epsilon_2}) \dots \phi(x_{i_m}^{\epsilon_m}) = 1$ and therefore

$$f(x_{i_1}^{\epsilon_1})f(x_{i_2}^{\epsilon_2}) \dots f(x_{i_m}^{\epsilon_m}) = 1$$

in $G$. This must hold for all groups $G$ and all functions $f$. If $a$ is not freely reducible to 1 and $f(x_i) \neq f(x_j)$ for $x_i \neq x_j$, $x_i, x_j \in X$, then we have a contradiction if $G$ is chosen, for example, as a free group itself, so that $f(a)$ is not a relation in $G$. Therefore, every relation with letters from $X$ in $F$ is freely reducible to 1, and therefore $F$ is a free group.

It remains to show that $X$ is a basis of $F$. If $X$ contained more elements than the rank of $F$ then there would be a relation $x_{i_1}^{\epsilon_1} x_{i_2}^{\epsilon_2} \dots x_{i_m}^{\epsilon_m} = 1$ among the elements of $X$ which does not have to hold in the image, i.e. $f(x_{i_1}^{\epsilon_1})f(x_{i_2}^{\epsilon_2}) \dots f(x_{i_m}^{\epsilon_m}) \neq 1$. We will prove in Theorem 5.16 that this is not possible.                                               □

**Theorem 5.13** *Every finitely generated group $G$ is a quotient of a free group.*

**Proof** We take a finite generating system $X$ of $G$ as a generating system of a free group $F$. According to Theorem 5.12, the identity on $X$, i.e. the mapping $f: X \to G$ which maps the elements of $X$ onto themselves, can be extended to a homomorphism $\phi: F \to G$. According to the 1st isomorphism theorem (Theorem 3.41), $G = F/\ker(\phi)$.                                               □

Let $F(X)$ be the free group over $X$. Let $G = \langle X \mid R \rangle$, where $X$ is a set of generators and $R$ is a set of relations in $X$. The *normal closure* of $R$ is the smallest normal subgroup in $F(X)$ which contains the elements of $R$, and is denoted by $\bar{R}$. By Lemma 3.38 the normal subgroup property for $\bar{R}$ is equivalent to $g\bar{R}g^{-1} \subset \bar{R}$ for all $g \in F(X)$. Since $\bar{R} \lhd F(X)$, for $r \in R$ we must also have $wrw^{-1} \subset \bar{R}$ for all words $w \in F(X)$. From these $wrw^{-1}$ we can form arbitrarily finite products without leaving the normal subgroup.

It follows that

$$\bar{R} \supset \left\{ \prod w_{i_j} r_{i_j}^{\epsilon_j} w_{i_j}^{-1} \right\},$$

where the product is formed over all possible finite products with $\epsilon_j = \pm 1, r_{i_j} \in R$ and $w_{i_j} \in F(X)$. But we also have

$$\bar{R} \subset \left\{ \prod w_{i_j} r_{i_j}^{\epsilon_j} w_{i_j}^{-1} \right\},$$

because $\bar{R}$ is the *smallest* normal subgroup in $F(X)$ that contains the elements of $R$. This gives us the following:

**Theorem 5.14**  *Let* $G = \langle X \mid R \rangle$. *Then:*

$$\bar{R} = \left\{ \prod w_{i_j} r_{i_j}^{\epsilon_j} w_{i_j}^{-1} \right\},$$

*where* $\epsilon_j = \pm 1$, $r_{i_j} \in R$ *and* $w_{i_j} \in F(X)$.

**Corollary 5.15**  *If* $G = \langle X \mid R \rangle$, *then* $G$ *is isomorphic to* $F(X)/\bar{R}$.

**Proof**  If $G = \langle X \mid R \rangle$, then precisely words of the form $w = \prod w_{i_j} r_{i_j}^{\epsilon_j} w_{i_j}^{-1}$ with $\epsilon_j = \pm 1$, $r_{i_j} \in R$ and $w_{i_j} \in F(X)$ can be represented as relations from $R$. We can conjugate relations and multiply these conjugates.                             $\square$

The following theorem of VAN DYCK describes in which cases a homomorphism can be found for a given presentation into a given group, namely when the relations hold in the image (i.e., images of relations are relations of the image group).

**Theorem 5.16**  *Let* $G = \langle X \mid R \rangle$ *and* $f : X \rightarrow H$ *be any function into a group* $H$. *Let* $\phi : F(X) \rightarrow H$ *be the associated homomorphism given by Theorem 5.12. If* $\phi(r) = \mathrm{id}$ *for all* $r \in R$, *then* $\phi$ *induces a homomorphism* $\psi : G \rightarrow H$ *with* $f(x) = \psi(x)$ *for all* $x \in X$.

**Proof**  From $\phi(r) = \mathrm{id}$ for all $r \in R$ it follows that $R \subset \ker(\phi)$. If $g \in G$, we write $g$ in the generators and their inverses:

$$g = x_{i_1}^{\epsilon_1} x_{i_2}^{\epsilon_2} \ldots x_{i_m}^{\epsilon_m}.$$

Here, each $x_{i_k} \in X$ and $\epsilon_k = \pm 1$. We define $\psi : G \rightarrow H$ by

$$\psi(g) = f(x_{i_1}^{\epsilon_1}) f(x_{i_2}^{\epsilon_2}) \ldots f(x_{i_m}^{\epsilon_m})$$

as in the proof of Theorem 5.12. The difference between two representations of $g$ in the generators is a relation, and this holds in the image, since $R \subset \ker(\phi)$.        $\square$

**Theorem 5.17**  *Every group has a group presentation.*

**Proof**  Choose for a group $G$ a generating system $X$ (for example all elements of $G$). From the universal mapping property we get a homomorphism $\phi : F(X) \rightarrow G$. The elements of $\ker(\phi)$ form the relations, so $G = \langle X \mid \ker(\phi) \rangle$.        $\square$

Caution: The resulting group presentation is by no means always finite. On the other hand, for example, finite groups always have a finite presentation: Take all group elements of a finite group $G$ as generators. For any two group elements $g, h \in G$ of a finite group $G$, calculate $g \cdot h$ and obtain a new group element $c = g \cdot h$. Then write all equations $g \cdot h = c$ as relations.

Let $g$ be an element of a free group $F$. We denote the *length* of $g$ by $|g|$, i.e., the sum of the absolute values of the exponents. For example, the word $a^3 b^{-2} a^6 b$ has length 12.

Let $F_n$ be the free group of rank $n$. It is clear that $F_n < F_m$ for $n < m$. Simply omit $m - n$ base elements from a basis of the group $F_m$ and obtain the group $F_n$.

Surprisingly, it also holds that $F_n < F_m$ for $n > m$. We show here $F_3 < F_2$ and consider the subset $H \subset F_2$ of elements with even length, i.e.,

$$H = \{g \in F_2 \mid |g| \text{ is even}\}.$$

**Theorem 5.18** *H is a subgroup of $F_2$. H is free of rank 3.*

**Proof** The identity element has length 0 and is therefore in $H$. The product of two elements of even length from $H$ is again even and thus also in $H$. The operation is therefore closed on $H$. The inverse of an element $g \in H$ has the same length as $g$ and is therefore also in $H$. Thus, $H < F_2$.

According to Theorem 5.11, subgroups of free groups are free, i.e., $H$ is a free group. If $F_2 = \langle a, b \rangle$, it can be proven that $\{a^2, ab, ab^{-1}\}$ forms a basis for $H$ (see [Mei08]), and thus $H$ has rank 3.                                                      □

**Theorem 5.19** $S_n = \langle x_1, \ldots, x_{n-1} \mid x_i^2, [x_i, x_j], \forall |i - j| > 1, (x_i x_{i+1})^3 \rangle.$

**Proof** Let $J_n$ be the group presented by

$$\langle x_1, \ldots, x_{n-1} \mid x_i^2, [x_i, x_j], \forall |i - j| > 1, (x_i x_{i+1})^3 \rangle.$$

We construct a homomorphism $\phi \colon J_n \to S_n$ by setting $\phi(x_i) = (i, i+1)$. According to Theorem 5.16, the homomorphism is well-defined if the relations in $J_n$ also hold in the image. It is clear that $\phi(x_i) = (i, i + 1)$ has order 2. The relations $x_i^2 = 1$ are therefore fulfilled in the image. $x_i$ commutes with $x_j$ for $|i - j| > 1$. In the image, this means that $(i, i + 1)$ must commute with $(j, j + 1)$, which is fulfilled for $|i - j| > 1$. Finally, we get $\phi(x_i x_{i+1}) = (i, i + 1, i + 2)$ has order 3, so the relations $(x_i x_{i+1})^3 = 1$ are fulfilled in the image.

We show that

$$S_n = \langle (1, 2), (2, 3), (3, 4), \ldots, (n - 1, n) \rangle$$

holds. Then the images of the generators of $J_n$ generate $S_n$, and we have proven that $\phi$ is surjective. Consider this: If $n$ people are sitting in a row, you can permute their positions arbitrarily by always letting only adjacent people swap their seats. If, for example, the person in seat 7 should go to seat 1, you can proceed as follows: The people in seats 7 and 6 swap their seats, then the people in seats 6 and 5, then 5 and 4, 4 and 3, 3 and 2 and finally 2 and 1. Thus, the person from seat 7 has moved to seat 1. Now look at which person should go to seat 2 and proceed in the same way. Proceed analogously with all other seats.

One can prove that $\phi$ is injective by showing that the order of $J_n$ is $n!$. To do this, consider the subgroup $H < J_n$ generated by $\{x_1, \ldots, x_{n-2}\}$. This satisfies all relations of $J_{n-1}$ and is therefore isomorphic to $J_{n-1}$. Inductively, we can assume that $J_{n-1}$ has order $(n-1)!$. Now show that the index of $H$ in $J_n$ is equal to $n$ by showing that there are only $n$ cosets. These are generated by the representatives

$$x_1 \ldots x_{n-1}, x_2 \ldots x_{n-1}, \ldots, x_{n-2}x_{n-1}, x_{n-1}$$

and an element of $H$. Therefore, $J_n$ has order $(n-1)! \cdot n = n!$ according to Corollary 3.16 and is therefore isomorphic to $S_n$.                                     □

**Example 5.20**  We analyze a group presentation with GAP. We consider the group

$$G = \langle x, y, z \mid x^2, y^2, z^2, xyz = yzx = zxy \rangle.$$

In GAP:

```
gap>  F := FreeGroup( "x", "y", "z");
<free group on the generators [ x, y, z ]>
gap> AssignGeneratorVariables(F);
#I  Assigned the global variables [ x, y, z ]
gap> G:=F/[x^2, y^2, z^2, x*y*z*x*z*y, x*y*z*y*x*z];
<fp group on the generators [ x, y, z ]>
gap> Order(G);
16
gap> Elements(G);
[ <identity ...>, x, y, z, (x*y)^2, x*y, x*z, y*x*y,
   y*z, x*y*x, x*z*x, x*y*z, y*x, z*x, z*y, x*z*y ]
```

So it is a group of order 16. It is not commutative because the commutator $[x, y] = (xy)^2$ appears as an element. Because of the relations $x^2, y^2$, both $x = x^{-1}$ and $y = y^{-1}$ hold.

From the relations, $zxz = yxy$ and $yzy = xzx$ follow immediately. Most words of length 4 can be transformed into shorter words. We show this using the example of $xzyx$. From the relation $xyz = yzx$ it follows that $zyx = xzy$ by inverting on both sides. Subsequently we get

$$xzyx = x \cdot zyx = x \cdot xzy = zy.$$

It is not hard to see that

$$(xy)^2 = (zy)^2 = (xz)^2 = (yx)^2 = (yz)^2 = (zx)^2.$$

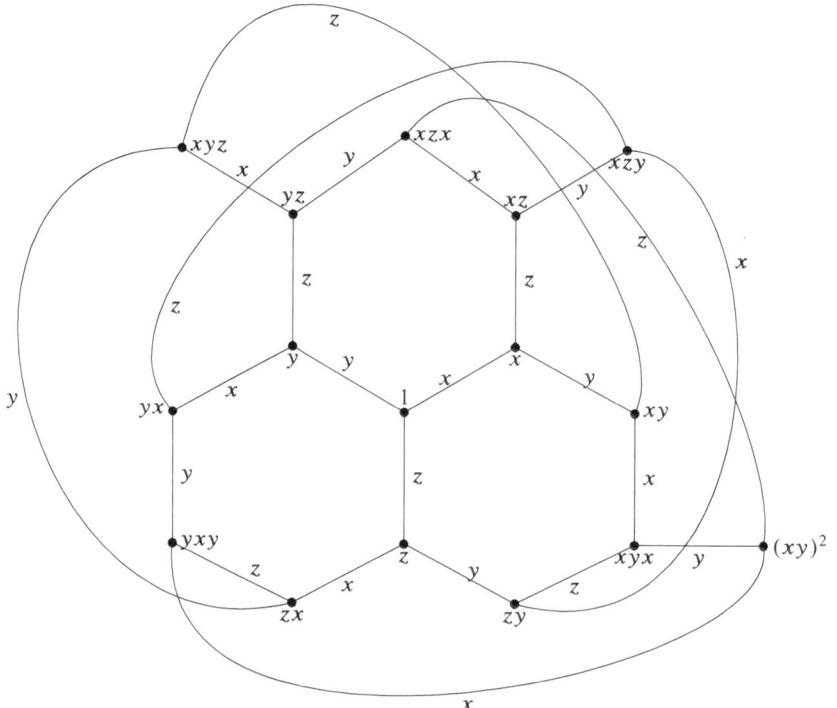

**Fig. 5.1**   Cayley graph of a group of order 16

Other words of length 4 can be reduced in length. With this information, the Cayley graph can be easily drawn (see Fig. 5.1).

With the help of the following *ping-pong lemma*, one can in some cases show that a given group is a free group (generalizations of this can be found in [Ce17]).

**Theorem 5.21** *Let $a$ and $b$ be generators of a group $G$ which acts on a set $X$. Let $X_a, X_b \subset X$ be disjoint subsets. Let $a^k(X_b) \subset X_a$ and $b^k(X_a) \subset X_b$ for all $k \in \mathbb{Z}, k \neq 0$. Then $G$ is isomorphic to the free group of rank 2.*

**Proof** We prove that no word of the form

$$a^{n_1}b^{n_2}a^{n_3}b^{n_4}\ldots a^{n_m}, \quad a^{n_1}b^{n_2}a^{n_3}b^{n_4}\ldots b^{n_m},$$

$$b^{n_1}a^{n_2}b^{n_3}a^{n_4}\ldots b^{n_m} \quad \text{or} \quad b^{n_1}a^{n_2}b^{n_3}a^{n_4}\ldots a^{n_m} \tag{5.2}$$

with all $n_i \neq 0$ and $m \geq 1$ is equal to the identity in the group. This is sufficient, because every element of $G$ must have one of these 4 forms, and there are therefore no relations in $G$.

The element $g = a^{n_1} b^{n_2} a^{n_3} b^{n_4} \ldots a^{n_m}$ satisfies $g(X_b) \subset X_a$, because: $a^{n_m}(X_b) \subset X_a$ and $b^{n_{m-1}} a^{n_m}(X_b) \subset b^{n_{m-1}}(X_a) \subset X_b$ etc. Because $g(X_b) \subset X_a$, $g$ is different from the identity element.

The other 3 forms of words from (5.2) are proven analogously. □

As in a game of table tennis, one jumps through an $a$-power from $X_b$ to $X_a$, and then the ball goes back with a $b$-power from $X_a$ to $X_b$ and then again with an $a$-power to $X_a$, etc.

**Example 5.22** Let $X$ be an infinite tree, and $a, b$ be bijective mappings from $X$ onto itself, each mapping a straight line (i.e., a path infinite in both directions) $l_a$ and $l_b$ onto itself by a translation. $l_a$ is then called the *axis* of $a$, and the operation of $a$ (the translation) is called *hyperbolic*. Furthermore, $l_a$ and $l_b$ should intersect in exactly one point.

An example of such a tree is the Cayley graph of the free group with 2 generators, depicted in Fig. F.2 of Appendix F. The line $l_a$ is the horizontal line through the vertex 1 and $l_b$ is the vertical line through the vertex 1. If you remove the vertex 1 from $X$, you get 4 components. $X_b$ consists of the upper and lower component, $X_a$ of the right and left component. It should be clear that $a^k(X_b) \subset X_a$ and $b^k(X_a) \subset X_b$ holds. $a^k$ shifts the vertex 1 to $X_a$ and thus everything that is attached to the vertex 1, namely also $X_b$.

### Exercises

1. Draw a part of the Cayley graph of the free group of rank 2.
2. Show that the group $H$ of Theorem 5.18 has index 2 in $F_2$.
3. Let $F$ be a free group and $g \in F, g \neq 1$. Show that for the centralizer $Z(g) = \langle g \rangle \cong \mathbb{Z}$ holds.
4. Draw the Cayley graph for the *generalized quaternion group*

$$\langle a, b \mid a^4 = b^2 = abab \rangle.$$

Proceed analogously to Example 5.20.

## 5.3 Tietze Transformations and Decidability

In Sect. 5.1 we described what it means to derive a relation from others. We clarify here what was said there:

Let $G = \langle X \mid R \rangle$, where $X$ is a set of generators and $R$ is a set of relations in $X$. If a word $r \in F(X)$ is in the normal closure $\bar{R}$, it can be derived from the other relations. So we can safely add it to the set of relations without changing the group. It follows that $G = \langle X \mid R, r \rangle$.

Let $w \in F(X)$ be any word, and $a$ be a new symbol that is not in $X$. Then $G = \langle X, a \mid R, a = w \rangle$ holds. With the new generator, we can't form more group elements than just with $X$. Any $a$ in any word can be replaced by $w$, and then we only have the words that can be formed already in $X$. We make the following definition.

**Definition 5.23** Let $G = \langle X \mid R \rangle$. The operations

$$\langle X \mid R \rangle \rightarrow \langle X \mid R, r \rangle \quad \text{for } r \in \bar{R} \tag{5.3}$$

$$\langle X \mid R \rangle \rightarrow \langle X, a \mid R, a = w \rangle \quad \text{for } w \in F(X) \tag{5.4}$$

and their inverses are called *Tietze transformations*.

With Tietze transformations, relations can be multiplied, since for relations $r_i \neq r_j$ we have

$$\langle X \mid r_1, \ldots, r_n \rangle \xrightarrow{(5.3)} \langle X \mid r_1, \ldots, r_n, r_i r_j \rangle \xrightarrow{(5.3)^{-1}}$$

$$\langle X \mid r_1, \ldots, r_{i-1}, r_i r_j, r_{i+1}, \ldots, r_n \rangle.$$

In the second step, $r_i$ is deleted. We are allowed to do this, because $r_i$ lies in the normal closure of $\{r_1, \ldots, r_{i-1}, r_i r_j, r_{i+1}, \ldots, r_n\}$. It holds that $r_i = r_i r_j \cdot r_j^{-1}$.

Similarly, it can easily be shown that the *conjugation* with any $w \in F(X)$, i.e., $r \rightarrow wrw^{-1}$ for a relation $r$, as well as the *inversion* of a relation $r \rightarrow r^{-1}$, can be realized with Tietze transformations.

TIETZE proved the following theorem in 1908:

**Theorem 5.24** *Two finite presentations present the same group if and only if they can be transformed into each other by a sequence of Tietze transformations.*

Before giving the proof, we consider two examples that show how complex Tietze transformations can be:

**Example 5.25** The presentation $\langle x, y \mid y^{-2}x^3, xy^{-1}x^{-1}y^{-1}xy \rangle$ presents the trivial group.

To prove this, we transform the trivial group presentation $\langle x, y \mid x, y \rangle$ into the above presentation by Tietze transformations. We invert the two relations to $R_1 = y^{-1}$ and $S_1 = x^{-1}$ and then let GAP do the work:

```
gap> F := FreeGroup( "x", "y");
<free group on the generators [ x, y ]>
gap> AssignGeneratorVariables(F);
#I  Assigned the global variables [ x, y ]
gap> R1:=y^-1;
y^-1
gap> S1:=x^-1;
x^-1
gap> R2:=R1*S1^-1;
y^-1*x
gap> S2:=S1*R2^-1;
```

```
x^-2*y
gap> R3:=S2^-1*R2;
y^-1*x^2*y^-1*x
gap> S3:=S2*R3^-1;
x^-2*y*x^-1*y*x^-2*y
```

For simplification, we now write the letter $u$ for $y^{-1}x^2$:

```
gap> u:=y^-1*x^2;;
gap> R4:=u*(S3*u*R3*u^-1)*u^-1;
x*y^-1*x^-1*y*x^-2*y
gap> S:=x^-1*R4*S3^-1*x;
y^-2*x^3
gap> R:=R4*u*S*u^-1;
x*y^-1*x^-1*y^-1*x*y
```

and we end with the given relations $S = y^{-2}x^3$, $R = xy^{-1}x^{-1}y^{-1}xy$.

**Example 5.26** The presentations

$$P = \langle x, y, z \mid xy = yz, yz = zx, zx = xy \rangle \quad \text{and} \quad Q = \langle a, b \mid a^3 = b^2 \rangle$$

present the same groups.

We replace in the first two relations of $P$ each $z$ by $xyx^{-1}$ and then eliminate with $(5.4)^{-1}$ the generator $z$ and the third relation. The first two relations become the same, up to conjugation, and we can delete one of them. We get $\langle x, y \mid xyx = yxy \rangle$. We add the generator $a$ and the relation $a = xy$. Then we replace each $y$ by $x^{-1}a$ and, with $(5.4)^{-1}$, we delete the generator $a$ and the relation $y = x^{-1}a$. This gives $\langle a, x \mid ax = x^{-1}a^2 \rangle$. If we introduce $b = ax$ and then delete $x$ and $x = a^{-1}b$, we finally get $Q$.

We play a little in GAP:

```
gap> F := FreeGroup( "x", "y", "z" );;
gap> x:=F.1;; y:=F.2;; z:=F.3;;
gap> G:=F/[x*y*z^-1*y^-1, y*z*x^-1*z^-1, z*x*y^-1*x^-1];
<fp group on the generators [ x, y, z ]>
gap> P := PresentationFpGroup( G );
<presentation with 3 gens and 3 rels of total length 12>
gap> SimplifyPresentation( P );
#I  there are 2 generators and 1 relator of total length 6
gap> TzPrintRelators(P);
#I  1. x*y*x*y^-1*x^-1*y^-1
```

With the command `SimplifyPresentation`, GAP tries to simplify a given presentation using Tietze transformations.

```
gap> TzSubstitute( P );
#I  substituting new generator _x4 defined by x*y
#I  eliminating y = x^-1*_x4
#I  there are 2 generators and 1 relator of total length 5
gap> TzPrintRelators(P);
#I  1. x*_x4^-2*x*_x4
```

With `TzSubstitute`, GAP tries Tietze transformations of type (5.4). For $a =\_x4$, the relation `x*_x4^-2*x*_x4` corresponds to the relation $ax = x^{-1}a^2$. Further attempts to simplify the presentation are fruitless. GAP does not consider $Q$ simpler than $\langle a, x \mid ax = x^{-1}a^2 \rangle$.

We now prove Theorem 5.24:

**Proof** If the presentation $P$ transforms into a presentation $P'$ through a sequence of Tietze transformations, then the corresponding groups are isomorphic, as we have seen at the beginning of this section. We prove the converse:

Let $P = \langle X \mid R \rangle$ and $Q = \langle Y \mid S \rangle$ be two presentations of the same group. We first modify $P$ so that the elements of $Y$ appear as generators. Since the elements of $X$ generate the group, we can write each element $y_i \in Y$ as a word $w_i$ in the elements of $X$:

$$y_i = w_i(X) \quad \forall y_i \in Y.$$

We modify $P$ with (5.4) to $P_1 = \langle X, Y \mid R, \ y_i = w_i(X), \ \forall y_i \in Y \rangle$.

The relations $S$ can be added to $P_1$ with (5.3). They must follow from $R$ since $P_1$ and $Q$ present the same group and $R$ is a set of defining relations. We thus obtain

$$P_2 = \langle X, Y \mid R, \ y_i = w_i(X), \ \forall y_i \in Y, \ S \rangle.$$

Since the elements of $Y$ also generate the group, we can conversely write each element $x_j \in X$ as a word $v_j$ in the elements of $Y$:

$$x_j = v_j(Y) \quad \forall x_j \in X.$$

The relations $x_j = v_j(Y)$ are relations in the group and can therefore be derived from the relations of $P_2$. We may therefore modify $P_2$ with (5.3) to

$$P_3 = \langle X, Y \mid R, \ y_i = w_i(X), \forall y_i \in Y, \ S, \ x_j = v_j(Y), \forall x_j \in X \rangle.$$

$P_3$ is symmetric, so it can also be obtained from $Q$ through Tietze transformations. We can therefore obtain $Q$ from $P_3$ through the corresponding inverse Tietze transformations.                                                                                          □

If the presentations $P$ and $Q$ are finite, then the sequence of Tietze transformations is finite. It follows that:

**Corollary 5.27** *Two finite presentations present the same group if and only if they can be transformed into each other through a finite sequence of Tietze transformations.*

Unfortunately, this theorem does not provide a concrete method to decide, for two given presentations, whether the corresponding groups are isomorphic or not. This decision problem is known in the literature as the *isomorphism problem*. It is an *undecidable* problem, i.e., there is no algorithm that takes two arbitrary finite presentations as input, stops after a finite number of steps, and outputs whether the presentations belong to isomorphic groups or not. The isomorphism problem is one of the three fundamental problems associated with group presentations that MAX DEHN formulated in 1911. It is not even solvable for the trivial group, i.e., for a given group presentation, there is no general method to decide whether it is a presentation of the trivial group.

The *word problem* is also one of the three fundamental problems. Let $P = \langle X \mid R \rangle$ be a presentation. We assume in the following that $P$ is a finite presentation, although the definitions and theorems partly also apply to infinite presentations. There are, however, some special difficulties with infinite presentations, which we can avoid in the finite case.

Let $w$ be a word in the generators $X$. The word problem asks whether $w$ is a relation in $P$ or not. In general, there is no method to decide the word problem. The first examples of finite presentations with undecidable word problem were discovered by NOVIKOV (1955), BOONE (1954) and BRITTON (1958) (for a detailed discussion, see [Rot95]). Of course, there are many groups in which the word problem is decidable (see Exercise 1.).

**Theorem 5.28** *If $P$ is a finite presentation of a group $G$ with decidable word problem, then every finite presentation of $G$ has decidable word problem.*

**Proof** Let $P = \langle X \mid R \rangle$ and $P' = \langle X' \mid R' \rangle$ be finite presentations of the same group, and assume $P$ has decidable word problem. Let $w'$ be a word in $P'$. Since $P$ and $P'$ are presentations of the same group, there is a function $\phi \colon X' \to F(X)$ that induces the isomorphism. Through $\phi$ we can write the word $w'$ in the generators $X$ and decide in $P$ whether the resulting word $w$ is a relation. Since $\phi$ induces an isomorphism, $w'$ is a relation if and only if $w$ is. We have solved the word problem in $P'$.                                                                                      □

This proof assumes the existence of an isomorphism. Finding it presupposes the solution of the isomorphism problem.

At first glance, it is surprising that the word problem is undecidable. We know that relations are exactly the words that lie in the normal closure of the defining relations, i.e., can be written as $\prod w_{i_j} r_{i_j}^{\epsilon_j} w_{i_j}^{-1}$, where $P = \langle X \mid r_1, \ldots, r_n \rangle$ and the words $w_j$ are from the free group generated by $X$. Since the set of generators and the defining relations are finite, $F(X)$ is countable and so is the set of conjugates of relations and their inverses. The set of relations is therefore countable.

So if a relation $w \in F(X)$ is given, one can prove it in finite time by enumerating the set of relations and running through the list until we find $w$. But if $w \in F(X)$ is an arbitrary word, and $w$ has not been found after a certain time, it is not clear whether it will ever appear in the list. If $w \neq 1$, the comparison algorithm will never stop! Even after a very long time, we still have no proof of the fact that $w \neq 1$.

For finite groups, the word problem is decidable. The *Todd–Coxeter algorithm*, found in 1936 by J. TODD and H. COXETER (see for example [Joh90]), solves the word problem for a finite presentation of a finite group and even gives the order of the group (the GAP command `Order(G)` uses this algorithm). It is easy to see that drawing the Cayley graph is possible only if the word problem is decidable. Before I draw the next vertex in the Cayley graph, I must be able to decide whether I have already drawn this vertex, i.e., whether my given vertex label in the group is equal to an already existing one.

Finally, the *conjugacy problem* for a given presentation $Q$ and two given words $u$, $v$ in the generators and their inverses asks whether $u$ is *conjugate* to $v$ in $Q$, i.e., whether there is a word $w$ in the generators of $Q$ with $u = wvw^{-1}$. For $v = id$ the conjugacy problem reduces to the word problem. Since the word problem is generally undecidable, so is the conjugacy problem.

If we have found a normal form for words in a presentation that can be determined algorithmically, this is already a solution to the word problem. We use this for the following theorem.

**Theorem 5.29** *The group* $D_n = \langle s, d \mid s^2, d^n, sdsd \rangle$ *has decidable word problem.*

*Proof* In the proof of Theorem 5.5 on page 97 we have given an algorithm which can determine for any word in $s$, $d$ and their inverses whether it is a relation or not.

$\square$

It is not hard to see that free groups have solvable word and conjugacy problem, as one can easily find a normal form for words in the free group. Two words of a free group $F$ describe the same group element in $F$ if and only if they are equal after free reduction.

### Exercises

1. Provide an algorithm for the word problem in finitely generated abelian groups. Solve the word problem for any group with a given Cayley graph.
2. Prove that inverting relations and conjugating relations with any word of the free group of the generators can be realized by Tietze transformations.
3. Prove that $\langle x, y \mid xy^2 = y^3x, \ yx^2 = x^3y \rangle$ is a presentation of the trivial group. (Hint: Use GAP.)
4. Let $G$ be the group generated by the reflection over the angle bisector and the translations of length 1 along the two coordinate axes in the plane. Provide a presentation of the group $G$. Then solve the word problem in $G$ by specifying a normal form for any word in the three generators.
5. Describe a solution to the conjugacy problem for free groups.

# Chapter 6
# Products of Groups

There are various ways to build "larger" groups from given groups. For instance, there are different ways to multiply groups together. We want to discuss three of these possibilities, direct products, free products, and semidirect products, in the following sections. With the help of the semidirect product, we can describe a frieze. At the end, we characterize the possible translation subgroups of symmetry groups in the Euclidean plane.

## 6.1 The Direct Product

**Definition 6.1** Let $A, B$ be nonempty sets. The *cartesian product* of $A$ and $B$ is the set of ordered pairs

$$A \times B = \{(a, b) \mid a \in A, b \in B\}.$$

**Definition 6.2** Let $G, H$ be groups. The *direct product* $(G \times H, \cdot)$ of $G$ and $H$ (sometimes also called the *direct sum* in an additively written group) is the group with the cartesian product of $G$ and $H$ as its set of elements and the component-wise operation

$$(g, h) \cdot (g', h') = (g \circ g', h * h') \quad \text{with } g, g' \in G, h, h' \in H.$$

Here, $\circ$ is the operation in $G$ and $*$ is the operation in $H$.

Obviously, $G \times H$ is a group: $(id, id)$ is the identity element. The inverse of $(g, h)$ is $(g^{-1}, h^{-1})$ if $g^{-1}$ is the inverse of $g$ and $h^{-1}$ is the inverse of $h$.

$G \times H$ contains $G$ and $H$ as subgroups, because $i_G : G \to G \times H$ defined by $i_G(g) = (g, id)$ is an *inclusion*, i.e., an injective homomorphism. The same applies

S. Rosebrock, *Visual Group Theory*, Springer Undergraduate Mathematics Series, https://doi.org/10.1007/978-3-662-69365-0_6

to $H$. In the group $G \times H$, we therefore often write $g$ for $(g, id)$ and $h$ for $(id, h)$. The elements of $G$ commute with the elements of $H$ in $G \times H$, because

$$g \cdot h = (g, id) \cdot (id, h) = (g, h) = (id, h) \cdot (g, id) = h \cdot g. \tag{6.1}$$

Therefore, $G$ is even a normal subgroup of $G \times H$, because with $g' \in G$ and $(g, h) \in G \times H$ we have

$$(g, h) \cdot g' \cdot (g, h)^{-1} = (g \cdot h) \cdot g' \cdot (g \cdot h)^{-1}$$
$$= g \cdot g' \cdot h \cdot h^{-1} \cdot g^{-1}$$
$$= g \cdot g' \cdot g^{-1} \in G.$$

The following theorem characterizes the direct product:

**Theorem 6.3** *$G$ is the direct product of its subgroups $U$ and $V$ if and only if:*

1. $G = U \cdot V$,
2. $U \triangleleft G$, $V \triangleleft G$,
3. $U \cap V = \{e\}$.

**Proof** We have already proven properties 1 and 2 for direct products. Property 3 can be seen as follows: From $(u, id) = (id, v)$ it follows that $u = id$ and $v = id$ for $u \in U$ and $v \in V$. Therefore, $U \cap V = (id, id) = id$.

We show the converse: Assume conditions 1, 2, and 3 hold. Because of 1, every element $g \in G$ can be represented by $g = uv$, $u \in U, v \in V$, just like in the cartesian product. We only need to check if the operation is formed just like in the direct product.

Because of 2, $u^{-1}v^{-1}u \in V$ and thus $u^{-1}v^{-1}uv \in V$. Similarly it implies that $v^{-1}uv \in U$ and therefore $u^{-1}v^{-1}uv \in U$. Therefore,

$$u^{-1}v^{-1}uv \in U \cap V,$$

and due to 3, $u^{-1}v^{-1}uv = 1$. This is equivalent to $uv = vu$, and in $G$ products are formed just like in $U \times V$:

$$uv \cdot u'v' = uu' \cdot vv'.$$

$\square$

If two groups are given by presentations, it is easy to write a presentation for their direct product. First we introduce some notation.

**Definition 6.4** The *commutator* of two group elements $g, h \in G$ is the element $ghg^{-1}h^{-1} \in G$ and is abbreviated as $[g, h]$.

If $U \subset G$ and $V \subset G$ are subsets of the group $G$, then accordingly

$$[U, V] = \{uvu^{-1}v^{-1} \mid u \in U, v \in V\}.$$

**Theorem 6.5** *If $\langle X \mid R \rangle$ is a presentation of the group $G$ and $\langle Y \mid S \rangle$ is a presentation of the group $H$, then $G \times H$ has the presentation $\langle X, Y \mid R, S, [X, Y] \rangle$.*

***Proof*** By item 1 of Theorem 6.3, every element of $G \times H$ can be written as a product of an element of $G$, which can be generated by elements of $X$, and an element of $H$, which can be generated by elements of $Y$. Therefore, $G \times H$ can be generated by the elements of $X, Y$.

The relations $R$ and $S$ still apply in $G \times H$. Above we have seen that the elements of $G$ commute with those of $H$ in $G \times H$. The relations $[X, Y]$ ensure this. If $g \in G$ and $h \in H$ are arbitrary elements, we can write them in the generators and their inverses:

$$g = x_{i_1}^{\epsilon_1} x_{i_2}^{\epsilon_2} \ldots x_{i_m}^{\epsilon_m}.$$

Here, each $x_{i_k} \in X$ and $\epsilon_k = \pm 1$. Similarly,

$$h = y_{j_1}^{\delta_1} y_{j_2}^{\delta_2} \ldots y_{j_n}^{\delta_n}$$

with $y_{j_k} \in Y$ and $\delta_k = \pm 1$. Now

$$g \cdot h = x_{i_1}^{\epsilon_1} x_{i_2}^{\epsilon_2} \ldots x_{i_m}^{\epsilon_m} \cdot y_{j_1}^{\delta_1} y_{j_2}^{\delta_2} \ldots y_{j_n}^{\delta_n} = y_{j_1}^{\delta_1} y_{j_2}^{\delta_2} \ldots y_{j_n}^{\delta_n} \cdot x_{i_1}^{\epsilon_1} x_{i_2}^{\epsilon_2} \ldots x_{i_m}^{\epsilon_m} = h \cdot g,$$

if the relations $[X, Y]$ hold.

We still need to show that no more relations are needed in $G \times H$. For this, we consider an arbitrary relation in $G \times H$:

$$w = x_{i_1}^{\epsilon_1} y_{j_1}^{\delta_1} x_{i_2}^{\epsilon_2} \ldots y_{j_n}^{\delta_n} = 1.$$

Through Tietze operations with the defining relations from $[X, Y]$ we can transform $w$ into

$$x_{i_1}^{\epsilon_1} x_{i_2}^{\epsilon_2} \ldots x_{i_n}^{\epsilon_n} y_{j_1}^{\delta_1} y_{j_2}^{\delta_2} \ldots y_{j_n}^{\delta_n}.$$

But this word is equal to $(1, 1)$, because $w$ is a relation, and thus the subword $x_{i_1}^{\epsilon_1} x_{i_2}^{\epsilon_2} \ldots x_{i_n}^{\epsilon_n}$ must be trivializable by the relations $R$ and the subword $y_{j_1}^{\delta_1} y_{j_2}^{\delta_2} \ldots y_{j_n}^{\delta_n}$ by the relations $S$. By using only the relations $R, S, [X, Y]$, we have transformed an arbitrary relation into the identity element, and thus the set $R, S, [X, Y]$ is a set of defining relations.                                                             □

A presentation of the group $(\mathbb{Z}_n, +)$ is $\langle x \mid x^n \rangle$. Indeed, the words are

$$1, x, x^2, \ldots, x^{n-1},$$

and the operation is defined by addition of exponents modulo $n$: $x^i x^j = x^{i+j \bmod n}$. From Theorem 6.5 we get the following:

**Corollary 6.6** $\mathbb{Z}_n \times \mathbb{Z}_m = \langle x, y \mid x^n, y^m, xy = yx \rangle$.

A presentation of $\mathbb{Z}_2 \times \mathbb{Z}_2$ is $\langle a, b \mid a^2, b^2, ab = ba \rangle$. This is the same presentation as in Example 5.7. Therefore, the symmetry group of the rhombus, i.e., the Klein four-group, is isomorphic to $\mathbb{Z}_2 \times \mathbb{Z}_2$.

The group $\mathbb{Z}_n$ is in GAP `CyclicGroup(n)`. We obtain the Klein four-group via

```
gap> G:=DirectProduct(CyclicGroup(2),CyclicGroup(2));
<pc group of size 4 with 2 generators>
gap> Elements(G);
[ <identity> of ..., f1, f2, f1*f2 ]
```

with the four elements $id$, two reflections and their product as a rotation by $180°$.

The automorphism group of the group $\mathbb{Z}_2 \times \mathbb{Z}_2$ is the symmetric group $S_3$. Indeed, the 3 elements that are not the identity can be permuted arbitrarily, and for each permutation, an automorphism of the Klein four-group is obtained.

**Example 6.7** The Cayley graph for the group $\mathbb{Z}_4 \times \mathbb{Z}_2$ generated by $(1, 0)$ and $(0, 1)$ is shown in Fig. 6.1.

Let $Q \in \mathbb{R}^3$ be a cuboid, in which exactly 2 opposite faces are squares. Let $\mathrm{Sym}^+(Q)$ be the group of orientation-preserving isometries of $Q$. We prove $\mathrm{Sym}^+(Q) \cong \mathbb{Z}_4 \times \mathbb{Z}_2$: Let $A$ be a fixed vertex of $Q$. It is clear that for each vertex $B \in Q$ there is one and only one element $g \in \mathrm{Sym}^+(Q)$ such that $g(A) = B$. Therefore, $\mathrm{Sym}^+(Q)$, like $\mathbb{Z}_4 \times \mathbb{Z}_2$, also has 8 elements, because $Q$ has 8 vertices.

**Fig. 6.1** A Cayley graph for the group $\mathbb{Z}_4 \times \mathbb{Z}_2$

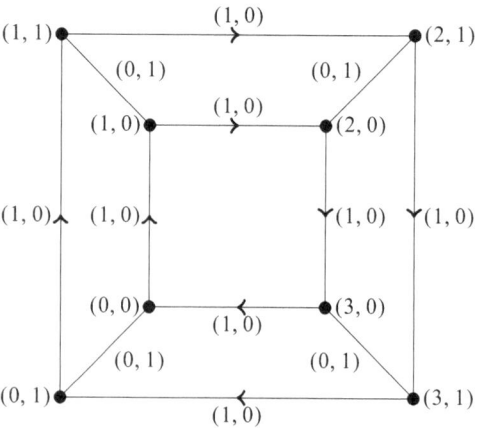

Let $d \in \mathrm{Sym}^+(Q)$ be the rotation by $90°$ about the axis that passes through the centers of the opposite square faces, and let $t$ be a rotation by $180°$ about an axis that passes through the centers of two opposite rectangle faces of $Q$. We obtain the isomorphism $\phi \colon \mathrm{Sym}^+(Q) \to \mathbb{Z}_4 \times \mathbb{Z}_2$ by setting $\phi(d) = (1, 0)$ and $\phi(t) = (0, 1)$.

Let $P$ be a cuboid in $\mathbb{R}^3$ without squares on the boundary and $\mathrm{Sym}(P)$ its symmetry group. We argue that $\mathrm{Sym}(P)$ is isomorphic to the group $\mathbb{Z}_2 \times \mathbb{Z}_2 \times \mathbb{Z}_2$ in a way analogous to the proof just given. Let $A$ be a fixed vertex of $P$. It is clear that for each vertex $B \in P$ there is one and only one element $g \in \mathrm{Sym}(P)$, such that $g(A) = B$. Therefore, $\mathrm{Sym}(P)$, like $\mathbb{Z}_2 \times \mathbb{Z}_2 \times \mathbb{Z}_2$, also has 8 elements. There are 3 different planes $E_1, E_2, E_3$, each passing through the midpoints of 4 parallel edges of $P$. Let $s_i$ be the reflection through the plane $E_i$.

We have $\mathrm{Sym}(P) = \langle s_1, s_2, s_3 \rangle$, as one can easily see. The isomorphism $\phi \colon \mathrm{Sym}(P) \to \mathbb{Z}_2 \times \mathbb{Z}_2 \times \mathbb{Z}_2$ is obtained by

$$\phi(s_1) = (1, 0, 0), \quad \phi(s_2) = (0, 1, 0), \quad \phi(s_3) = (0, 0, 1).$$

The dihedral group $D_6$ is isomorphic to $S_3 \times \mathbb{Z}_2$. GAP gives us the isomorphism. We consider a regular hexagon with the vertices numbered $1, \ldots, 6$. We generate the group $D_6$ by a reflection over the line through the vertices $1, 4$ and the rotation $d$ by $60°$.

```
gap> G:=DirectProduct(SymmetricGroup(3),CyclicGroup(2));
<group of size 12 with 3 generators>
gap> D6:=Group((6,2)(3,5),(1,2,3,4,5,6));
Group([ (2,6)(3,5), (1,2,3,4,5,6) ])
gap> IsomorphismGroups(G,D6);
[ DirectProductElement( [ (1,2,3), <identity> of ... ] ),
  DirectProductElement( [ (1,2), <identity> of ... ] ),
  DirectProductElement( [ (), f1 ] ) ] ->
  [ (1,3,5)(2,4,6), (1,3)(4,6), (1,4)(2,5)(3,6) ]
```

In the isomorphism $\phi \colon S_3 \times \mathbb{Z}_2 \to D_6$, the element $((1, 2, 3), 0)$ is mapped to $d^4$, $((1, 2), 0)$ to the reflection over the line through the vertices $2, 5$ and $((), 1)$ to the point reflection $d^3$.

**Example 6.8**   Consider the two groups with 4 elements $\mathbb{Z}_4$ and $\mathbb{Z}_2 \times \mathbb{Z}_2$. Both have normal subgroups that are isomorphic to $\mathbb{Z}_2$, and the respective factor groups are also isomorphic to $\mathbb{Z}_2$. So there are different groups with isomorphic normal subgroups, such that the factor groups are also isomorphic.

**Exercises**

1. Let $Z$ be a partition of the plane into equal-sized squares and $G_{(4,4)}$ the associated symmetry group (see Exercise 2. of Sect. 4.6).

   (a) Provide a presentation for the group $G_{(4,4)}$.
   (b) Let $T$ be the subgroup of translations of $G_{(4,4)}$. Show that $T \cong \mathbb{Z} \times \mathbb{Z}$ holds.

2. Let $g \in G$ be an element of order $n$ and $h \in H$ an element of order $m$. What is the order of $(g, h) \in G \times H$?

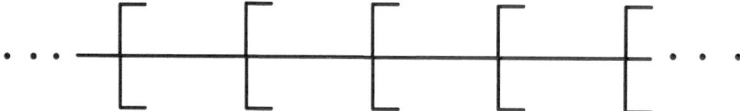

**Fig. 6.2**  Frieze

3. Prove that the symmetry group of the frieze of Fig. 6.2 is isomorphic to $\mathbb{Z} \times \mathbb{Z}_2$, and provide a presentation of the group.
4. Define an isomorphism $\phi \colon \mathbb{Z}_6 \to \mathbb{Z}_3 \times \mathbb{Z}_2$, and prove that it is an isomorphism. Generalize your observation to $\psi \colon \mathbb{Z}_n \to \mathbb{Z}_p \times \mathbb{Z}_q$, where $p$ and $q$ are *coprime* (i.e., the greatest common divisor of $p$ and $q$ is 1) and $n = p \cdot q$.
5. (a) We define an operation on the set $M$ of those natural numbers which only have prime factors $2, 3, 5$ and $7$. Also assume $1 \in M$. $a \diamond b$ is the ordinary product of $a$ and $b$, where common prime factors are canceled. For example $12 \diamond 42 = (2 \cdot 2 \cdot 3) \diamond (2 \cdot 3 \cdot 7) = 2 \cdot 7 = 14$. Why is $(M, \diamond)$ not a group?
   (b) Modify $M$ so that $(M, \diamond)$ becomes a group.
   (c) To which known group is $(M, \diamond)$ isomorphic?
6. Prove that $S_n$ for $n \geq 3$ is not isomorphic to the group $A_n \times \mathbb{Z}_2$ by using the fact that the center of $S_n$ is trivial (see Theorem 4.45), but that of $A_n \times \mathbb{Z}_2$ is not.

## 6.2   The Free Product

**Definition 6.9**  Let $G$ be a group given by the presentation $\langle X \mid R \rangle$, and $H$ be a group given by the presentation $\langle Y \mid S \rangle$. The *free product* $G \ast H$ of the groups $G$ and $H$ has the presentation $\langle X, Y \mid R, S \rangle$. The groups $G$ and $H$ are then called *free factors* of $G \ast H$.

This definition only makes sense if the free product is independent of the concrete presentations:

**Theorem 6.10**  *The free product $G \ast H$ depends only on the groups $G$ and $H$.*

**Proof**  Let $\langle X' \mid R' \rangle$ be another presentation of $G$ and $\langle Y' \mid S' \rangle$ another presentation of $H$. The isomorphism $\phi_G$ from $\langle X' \mid R' \rangle$ to $\langle X \mid R \rangle$ and the isomorphism $\phi_H$ from $\langle Y' \mid S' \rangle$ to $\langle Y \mid S \rangle$ lead, according to Theorem 5.16, to a homomorphism $\phi$ from $\langle X', Y' \mid R', S' \rangle$ to $\langle X, Y \mid R, S \rangle$, because the defining relations $R', S'$ are mapped to relations in $\langle X, Y \mid R, S \rangle$. $\phi$ can be inverted by the inverses of $\phi_G$ and $\phi_H$ and is thus itself an isomorphism.                                            □

A *frieze* is a figure in the plane with translation subgroup $\mathbb{Z}$. We have already encountered friezes, for example in Fig. 1.3.

**Example 6.11**  The frieze of Fig. 6.3 has as its symmetry group $D_\infty = \mathbb{Z}_2 \ast \mathbb{Z}_2$, the *infinite dihedral group*.

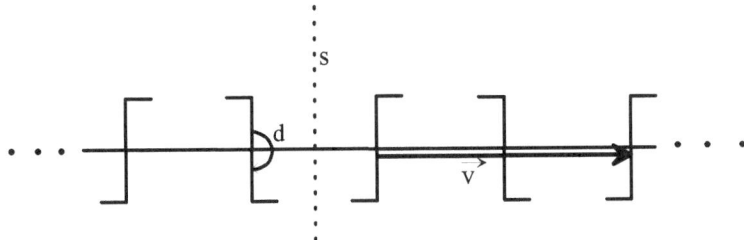

**Fig. 6.3** Band ornament

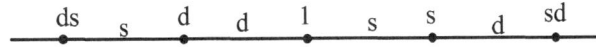

**Fig. 6.4** Cayley graph for $\mathbb{Z}_2 * \mathbb{Z}_2$

To see this, observe that $D_\infty$ is generated by the reflection $s$, the rotation $d$ by $180°$, and the translation $\tau$ along $\vec{v}$ (see Fig. 6.3). One obtains the reflections over parallels to $s$ through $\tau^n s$ for $n \in \mathbb{Z}$, and the further $180°$ rotations through $\tau^m d$ for $m \in \mathbb{Z}$. The glide reflection by the vector $\vec{v}/2$ is obtained through $sd$. If one follows one of the small, horizontal lines of Fig. 6.3 under these isometries, one can map this line to any other horizontal line. Therefore, $s, d, \tau$ generate $D_\infty$.

If one performs the glide reflection twice, one obtains the translation $\tau$. Therefore, $\tau = (sd)^2$ is a relation in $D_\infty$. We obtain the following presentation:

$$D_\infty = \langle s, d, \tau \mid s^2, d^2, s\tau s = \tau^{-1}, d\tau d = \tau^{-1}, \tau = (sd)^2 \rangle. \tag{6.2}$$

Since $s$ is a reflection, the relation $s^2$ applies. Similarly, a rotation by $180°$ has order 2 and leads to the relation $d^2$. The relations $s\tau s = \tau^{-1}, d\tau d = \tau^{-1}$ are easily seen by looking at the figure.

We do not need more relations in (6.2). In order to see this, we first transform this presentation through Tietze transformations into another presentation: If one replaces $\tau$ by $(sd)^2$ in $s\tau s = \tau^{-1}$ and $d\tau d = \tau^{-1}$, these two relations with the help of $s^2 = 1$ and $d^2 = 1$ become trivial relations $s \cdot sdsd \cdot s = dsds$ and $d \cdot sdsd \cdot d = dsds$ and can therefore be omitted. Then we cross out $\tau$ and the relation $\tau = (sd)^2$ with $(5.4)^{-1}$ and thus obtain

$$D_\infty = \langle s, d \mid s^2, d^2 \rangle. \tag{6.3}$$

A part of the Cayley graph corresponding to the generators $s, d$ is depicted in Fig. 6.4. This is not the same Cayley graph as for $(\mathbb{Z}, +)$. Here we have replaced each pair of oriented edges between the same points with an unoriented edge, because $d$ and $s$ both have order 2.

From the Cayley graph we see that the elements of (6.3) are exactly

$$\{s, d, sd, ds, sds, dsd, (sd)^2, (ds)^2, (sd)^2 s, (ds)^2 d, \ldots\}.$$

In Fig. 6.3 we can see that these isometries are all different. Therefore, we do not need any additional relations in (6.3) and thus also in (6.2). $D_\infty$ is the symmetry group of the graph consisting of a path which extends infinitely in both directions.

If $G = H_1 * \cdots * H_n$, and $g \in G$ is nontrivial, then $g$ has a representation $g = g_1 \ldots g_m$, where each $g_i$ is a nontrivial element of one $H_j$, and $g_{i-1}$ and $g_i$ lie in different factors $H_j$ and $H_k$. This is because, according to the definition of the free product, we can write the element $g$ in the generators of the factors.

**Theorem 6.12** *If the groups $G, H$ are each finitely generated with solvable word problem, then $G * H$ is finitely generated with solvable word problem.*

**Proof** We write a word $w$ in the generators as

$$w = g_1 h_1 g_2 h_2 \ldots g_n h_n,$$

where $g_i \in G, h_i \in H$ and $h_i \neq 1$ for $1 \leq i < m$ and $g_i \neq 1$ for $1 < i \leq m$. We can always achieve this by omitting factors of $G$ or $H$ that describe the trivial element.

Now $w = 1$ in $G * H$ if and only if $m = 1, g_1 = 1$ and $h_1 = 1$.                    □

**Exercises**

1. Prove that the free product of two nontrivial groups always has infinite order.
2. Prove that the free group of rank 2 is isomorphic to $\mathbb{Z} * \mathbb{Z}$.

## 6.3   The Semidirect Product

If $G = A \times B$, then $G/A = B$. The converse, however, is not true. If in general $G/N = H$, then $G$ is called an *extension* of $N$ by $H$. One type of extension, the direct product, was the subject of Sect. 6.1. The semidirect product is another type of extension.

We know from Sect. 3.5 that every isometry of the plane can be represented by an isometry that leaves the origin fixed, followed by a translation, i.e. $\mathcal{E} = \mathcal{T} O_2$. According to Theorem 3.44, the subgroup of translations is a normal subgroup in $\mathcal{E}$, and there is a surjective homomorphism $\phi \colon \mathcal{E} \to O_2$ with kernel $\mathcal{T}$, so $\mathcal{E}/\mathcal{T} \cong O_2$.

Thus, $\mathcal{E}$ is an extension of $\mathcal{T}$ by $O_2$. We have $O_2 < \mathcal{E}$ and $\mathcal{T} \triangleleft \mathcal{E}$. In addition, there is an injective homomorphism $\psi \colon O_2 \to \mathcal{E}$, simply by considering the elements of $O_2$ as elements of $\mathcal{E}$. We have $\phi\psi(g) = g$ for all $g \in O_2$. This is an example of a semidirect product $\mathcal{E} = \mathcal{T} \rtimes O_2$:

**Definition 6.13** An extension $G$ of $N$ by $H$ is called a *semidirect product* if there is a surjective homomorphism $\phi: G \to H$ with $\ker(\phi) = N$ and a homomorphism $\psi: H \to G$ with $\phi\psi(h) = h$, $\forall h \in H$. We denote this by $G = N \rtimes H$.

If $G = N \rtimes H$, then the associated homomorphism $\psi: H \to G$ is injective, because from $\psi(h) = \psi(h')$ it follows that $\phi\psi(h) = \phi\psi(h')$ and therefore $h = \phi\psi(h) = \phi\psi(h') = h'$.

There is a characterization of the semidirect product analogous to Theorem 6.3. To prove it, we need:

**Theorem 6.14 (2nd Isomorphism Theorem)** *If $N$ is normal in $G$ and $H$ is any subgroup of $G$, then*

$$(N \cdot H)/N \cong H/(H \cap N).$$

Here, $N \cdot H = \{n \cdot h \mid h \in H, n \in N\}$, as usual.

***Proof*** The mapping $h \to Nh$, which assigns to each $h \in H$ the coset $Nh$, is a surjective homomorphism $\phi: H \to (N \cdot H)/N$. The kernel of $\phi$ consists of the elements of $H$ that also lie in $N$, i.e., $H \cap N$. The result follows from the first isomorphism theorem. $\square$

While we're at it, let's prove this as well:

**Theorem 6.15 (3rd Isomorphism Theorem)** *Let $H$, $N$ be normal in $G$, and $H$ be contained in $N$. Then $N/H$ is a normal subgroup of $G/H$ and*

$$(G/H)/(N/H) \cong G/N.$$

***Proof*** We define a mapping $\phi: G/H \to G/N$ by $\phi(gH) = gN$. This mapping is a homomorphism, because

$$\phi(gH)\phi(g'H) = gNg'N = gg'N = \phi(gg'H).$$

This homomorphism is surjective, because for every coset $gN$ there is a preimage $gH$. The coset $gH$ belongs to the kernel of $\phi$ if and only if $gN = N$, i.e., when $g \in N$. Thus, $\ker(\phi) = N/H$. According to Theorem 3.39, $N/H$ is a normal subgroup of $G/H$, and according to Theorem 3.41, $(G/H)/(N/H)$ is isomorphic to $G/N$. $\square$

**Example 6.16** We choose $G = \mathbb{Z}$, $N = 3\mathbb{Z}$ and $H = 9\mathbb{Z}$. Then $H \subset N$, and $N$, $H$ are normal subgroups of $G$, because $G$ is abelian. According to Theorem 6.15, it follows that

$$\mathbb{Z}_3 = \mathbb{Z}/3\mathbb{Z} = (\mathbb{Z}/9\mathbb{Z})/(3\mathbb{Z}/9\mathbb{Z}) = (\mathbb{Z}/9\mathbb{Z})/(\mathbb{Z}/3\mathbb{Z}) = \mathbb{Z}_9/\mathbb{Z}_3.$$

Next is the promised characterization of the semidirect product:

**Theorem 6.17**  *G is the semidirect product of its subgroups N and H if and only if:*

1. $G = N \cdot H$,
2. $N \lhd G$, $H < G$,
3. $N \cap H = \{e\}$.

**Proof**  If $G$ is the semidirect product of $N$ with $H$, then $G = N \cdot H$. Since $N$ is the kernel of a homomorphism, $N$ is normal in $G$. By definition, $N$ and $H$ only have the identity in common.

Conversely, assume conditions 1 to 3 are fulfilled. From $G = N \cdot H$ and the second isomorphism theorem, it follows that

$$G/N \cong (N \cdot H)/N \cong H/(H \cap N).$$

From the third condition, $H/(H \cap N) = H/\{id\} = H$, so that $G/N = H$. Therefore, $G$ is an extension of $N$ by $H$. The required homomorphism $\psi$ is the inclusion $\psi : H \to G$.                                                                    □

**Example 6.18**  We claim $S_n = A_n \rtimes K$ where $K = \langle (1, 2) \rangle \cong \mathbb{Z}_2$. First observe that $S_n = A_n \cdot K$, because if $\sigma \in A_n$ we can write $\sigma$ as $\sigma \cdot ()$ and if $\sigma \notin A_n$ we have $\sigma(1, 2) \in A_n$ and we can write $\sigma$ as $\sigma(1, 2) \cdot (1, 2)$. In both cases $\sigma$ is written as an element of $A_n$ multiplied by an element of $K$. This proves item 1 of Theorem 6.17.

$A_n$ and $K$ are subgroups of $S_n$ and $A_n$ is normal because it has index 2 in $S_n$ (see Exercise 3. of Sect. 3.4). $A_n \cap K = \{()\}$ since $(1, 2)$ is not an even permutation (see Sect. 4.1).

So Theorem 6.17 tells us that $S_n = A_n \rtimes K$. This semidirect product is different from the direct product $A_n \times K$, because $S_n$ has trivial center (see Theorem 4.45) and the center of $A_n \times K$ contains the element $((), (1, 2))$.

If a group $G$ can be written as $U \cdot V$ for subgroups $U$ and $V$, it will not necessarily be a direct or semidirect product. In Exercise 6. of Sect. 4.6, an example of a product was introduced that is not a semidirect product, as both factors are not normal subgroups.

Given two groups $H$ and $N$, their direct product is uniquely determined. This is not the case with the semidirect product. Every element of $G = N \rtimes H$ can be uniquely written as a product $nh$ and, because $N$ is a normal subgroup, as $hn$ for $h \in H$ and $n \in N$. For two elements $h_1 n_1, h_2 n_2$ from $N \rtimes H$, it follows that

$$h_1 n_1 \cdot h_2 n_2 = h_1 h_2 \cdot h_2^{-1} n_1 h_2 \cdot n_2 = h_1 h_2 n_1^{h_2} n_2.$$

In this case, the conjugation of $n_1$ with $h_2$ was set to $h_2^{-1} n_1 h_2 = n_1^{h_2}$. Since $N$ is normal in $G$, $n_1^{h_2} \in N$. In fact, for a fixed $g \in G$, the mapping $\alpha_g : N \to N$, defined by $n \to n^g = g^{-1} n g$, is an automorphism. Therefore, for each $h \in H$, there is an automorphism $\alpha_h$ of $N$, or, in other words, for a semidirect product $G = N \rtimes H$, there is a homomorphism from $H$ to the automorphism group of $N$, defined by $h \rightarrow \alpha_h$.

Conversely, if only the groups $N, H$ and a homomorphism from $H$ to the automorphism group of $N$ are given, a semidirect product is defined by this, because then every product $h_1 n_1 \cdot h_2 n_2$ is determined. If all automorphisms $n \rightarrow n^h$ are the identity, i.e., $n = n^h$ for all $n \in N$ and all $h \in H$, then the corresponding semidirect product is a direct product, and $H$ is a normal subgroup in the resulting product.

In Example 6.18 the homomorphism from $K$ to the automorphism group of $A_n$

$$v: \langle (1,2) \rangle \rightarrow \text{Aut}(A_n) \quad \text{is given by} \quad v((1,2))(p) = (1,2)p(1,2)$$

for a permutation $p \in A_n$ (remember $(1,2)^{-1} = (1,2)$).

We consider the dihedral group $D_n$. In $D_n$, we show that the relation

$$d^i s = s d^{n-i} \tag{6.4}$$

holds, where $d$ is a rotation by $360/n$ degrees and $s$ is a reflection. $d^i s$ is a reflection and therefore has order 2. So $d^i s d^i s = id$, and thus $d^i s d^i = s$ and after multiplying by $d^{n-i}$ from the left, Eq. (6.4) follows.

We have $D_n = \mathbb{Z}_n \cdot \mathbb{Z}_2$, because we can represent every element of $D_n$ as a pair $(d^k, s^\epsilon)$, or in the shorter form $d^k s^\epsilon$ ($0 \le k < n$ and $s \in \{0, 1\}$), where $d$ is the rotation by $360/n$ degrees and $s$ is the reflection. However, $D_n$ is not isomorphic to $\mathbb{Z}_n \times \mathbb{Z}_2$, because $\mathbb{Z}_2$ is not a normal subgroup in $D_n$. Left and right cosets are different: $d\{id, s\} \ne \{id, s\}d$, because $ds \ne sd$.

We compose two elements of $D_n$. In the first element, we do not take the identity element in the $\mathbb{Z}_2$ factor: $d^k s \cdot d^m s^\delta$. If $D_n$ were the direct product of $\mathbb{Z}_n$ and $\mathbb{Z}_2$, then the result would be $d^{k+m} s^{1+\delta}$. But this is not the case, instead we have to push $d^m$ past $s^\epsilon$ using Eq. (6.4):

$$d^k s \cdot d^m s^\delta = d^k d^{n-m} s^{1+\delta}.$$

The homomorphism $\phi: \mathbb{Z}_2 \rightarrow \text{Aut}(\mathbb{Z}_n)$ described above is obtained by mapping $s$ to the automorphism that maps each element to its inverse in $\mathbb{Z}_n$.

In the case when $N = \mathbb{Z}_k = \langle y \mid y^k \rangle$ and $H = \mathbb{Z}_m = \langle x \mid x^m \rangle$, we want to write down presentations for semidirect products. An automorphism of $N$ must map $y$ to an element of order $k$, i.e., to a $y^l$, so that $l$ and $k$ are coprime (compare Exercise 7. of Sect. 3.3). The conjugation of $y$ with $x$ gives this element, so that

$$\mathbb{Z}_k \rtimes \mathbb{Z}_m = \langle x, y \mid x^m, y^k, x^{-1} y x = y^l \rangle,$$

if $k$ and $l$ are coprime and $l^m \equiv 1 \bmod k$. The last condition is necessary because: The mapping from $H$ to the automorphism group of $N$ must be a homomorphism, and therefore the image of $x$ must have the same order as $x$, namely $m$. $y^m$ is mapped to $y^{l^m}$, and this must be $y$, so $l^m \equiv 1 \bmod k$ must hold.

In GAP, we consider $\mathbb{Z}_7 \rtimes \mathbb{Z}_3$. GroupHomomorphismByImages generates a homomorphism, in our case from $\mathbb{Z}_3$ into the automorphism group of $\mathbb{Z}_7$, where the generating element of $\mathbb{Z}_3$ is mapped to the automorphism of $\mathbb{Z}_7$ that doubles each element. $1 \in \mathbb{Z}_3$ doubles, $2 \in \mathbb{Z}_3$ quadruples, and $3 = 0 \in \mathbb{Z}_3$ octuples, which corresponds to 1 in $\mathbb{Z}_7$. Thus, 0 is assigned to the trivial automorphism, giving us a homomorphism $\phi \colon \mathbb{Z}_3 \to \mathrm{Aut}(\mathbb{Z}_7)$. The command SemidirectProduct then constructs the associated semidirect product.

```
gap> N:=CyclicGroup(7);; H:=CyclicGroup(3);;
gap> AutN:=AutomorphismGroup(N);
<group with 1 generators>
gap> el:=Elements(AutN);
[ IdentityMapping( <pc group of size 7 with 1 generators> ),
  Pcgs([ f1 ]) -> [ f1^2 ], [ f1 ] -> [ f1^3 ],
  Pcgs([ f1 ]) -> [ f1^4 ],
  Pcgs([ f1 ]) -> [ f1^5 ], Pcgs([ f1 ]) -> [ f1^6 ] ]
gap> hom := GroupHomomorphismByImages(H, AutN,
> GeneratorsOfGroup(H),[el[2]]);
[ f1 ] -> [ Pcgs([ f1 ]) -> [ f1^2 ] ]
gap> p:=SemidirectProduct(H,hom,N);
<pc group of size 21 with 2 generators>
gap> Elements(p);
[ <identity> of ..., f1, f2, f1^2, f1*f2, f2^2, f1^2*f2,
  f1*f2^2, f2^3, f1^2*f2^2, f1*f2^3, f2^4, f1^2*f2^3,
  f1*f2^4, f2^5, f1^2*f2^4, f1*f2^5, f2^6, f1^2*f2^5,
  f1*f2^6, f1^2*f2^6 ]
```

For another homomorphism to $\mathrm{Aut}(N)$, we get the same semidirect product in this case:

```
gap> hom2 := GroupHomomorphismByImages(H, AutN,
> GeneratorsOfGroup(H),[el[4]]);
[ f1 ] -> [ Pcgs([ f1 ]) -> [ f1^4 ] ]
gap> p2:=SemidirectProduct(H,hom2,N);;
gap> IsomorphismGroups(p,p2);
[ f1, f2 ] -> [ f1^2*f2^4, f2^4 ]
```

IsomorphismGroups indicates an isomorphism between two groups if they are isomorphic and returns fail otherwise. The mapping that maps the generator of $H$ to the automorphism f1 -> f1^3 in $N$ is not a homomorphism because $3^3 \not\equiv 1 \bmod 7$. This contradicts the above condition: $l^m \equiv 1 \bmod k$.

Finally, we form the semidirect product $\mathbb{Z}_7 \rtimes \mathbb{Z}_3$ over the trivial automorphism, i.e., the direct product. This turns out to be different from the previously constructed semidirect product.

```
gap> hom3 := GroupHomomorphismByImages(H, AutN,
>  GeneratorsOfGroup(H),[el[1]]);
[ f1 ] -> [ IdentityMapping( Group( [ f1 ] ) ) ]
gap> p3:=SemidirectProduct(H,hom3,N);;
gap> IsomorphismGroups(p,p3);
fail
```

**Theorem 6.19** $D_n \cong \mathbb{Z}_n \rtimes \mathbb{Z}_2$.

***Proof*** According to Theorem 5.5 on page 97, we have $D_n = \langle s, d \mid s^2, d^n, sdsd \rangle$. We rewrite the last relation as $sds^{-1} = d^{n-1}$. $n$ is coprime to $n-1$ and $(n-1)^2 \equiv 1$ mod $n$. Therefore,

$$\mathbb{Z}_n \rtimes \mathbb{Z}_2 = \langle s, d \mid s^2, d^n, sds^{-1} = d^{n-1} \rangle$$

according to the above.                                                          □

**Exercises**

1. Let $G$ be a group of order $n$ and $H$ a group of order $m$. What are the orders of $G \times H$ and $G \rtimes H$?
2. Write the stabilizer $\mathcal{E}(l)$ of a line $l$ in $\mathcal{E}$ as a semidirect product of its translation subgroup with the subgroup of $\mathcal{E}(l)$ which leaves a point $P \in l$ fixed.

## 6.4  Discontinuous Groups and Translations

There is a significant difference between the symmetry group of a circle and a regular $n$-gon in that a circle allows arbitrarily small rotations and the regular $n$-gon does not. Similarly, the symmetry group of a straight line allows arbitrarily small translations, unlike friezes, which only allow translations by vectors that do not fall below a certain length (their translation subgroup is always $\mathbb{Z}$ by definition). We summarize this difference in a definition:

**Definition 6.20**  A group $G$ of isometries in $\mathbb{R}^2$ operates *discontinuously* if for every point $P \in \mathbb{R}^2$ there is a disc $D$ with center $P$ that contains no image of $P$ under $G$ except $P$, i.e., $\forall g \in G$ with $g \neq id$ it holds that $g(P) \notin D$ or $g(P) = P$.

It is not difficult to see that this definition is equivalent to the following: A group $G$ of isometries in $\mathbb{R}^2$ is called *discontinuous* if for every point $P \in \mathbb{R}^2$ and for every disc $D \subset \mathbb{R}^2$ the orbit of $P$ has only finitely many points in $D$ (see Exercise 1.).

If a group is discontinuous, then of course every one of its subgroups is discontinuous. Recall that $\mathcal{T} < \mathcal{E}$ is the subgroup of translations in the group $\mathcal{E}$ of the plane.

**Theorem 6.21**  *Let $T < \mathcal{T}$ be discontinuous. Then $T$ is either the trivial group, isomorphic to $\mathbb{Z}$ or isomorphic to $\mathbb{Z} \times \mathbb{Z}$.*

***Proof*** We assume that $T$ is not the trivial group. Let $P$ be an arbitrary point of the Euclidean plane. We choose a disk with center $P$ large enough that points from the orbit $TP$ are included. Because $T$ is discontinuous, there are only finitely many. Let $\tau \in T$ be the translation such that $\tau(P)$ is one of the elements closest to $P$ from the orbit $TP$. Let $l$ be the line through $P$ and $\tau(P)$, and $T(l)$ be its stabilizer in $T$. We show that $T(l)$ is isomorphic to $\mathbb{Z}$ with generator $\tau$. Otherwise there would be a translation $\alpha \in T(l)$, such that $\alpha(P)$ lies in the open interval $]\tau^k(P), \tau^{k+1}(P)[$.

**Fig. 6.5** Fundamental region

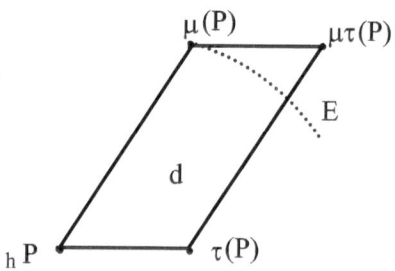

Then, however, the point $\tau^{-k}\alpha(P)$ would be closer to $P$ than $\tau(P)$, contradicting the definition of $\tau$.

If $T = T(l)$, then $T$ is isomorphic to $\mathbb{Z}$. If $T \neq T(l)$, then among all translations that are not from $T(l)$, there is a translation $\mu$ such that $\mu(P)$ has the smallest distance from $P$. This distance is at least as large as the distance from $\tau(P)$ to $P$ by the definition of $\tau$.

Let $d$ be the parallelogram with vertices $P, \mu(P), \mu\tau(P), \tau(P)$ (see Fig. 6.5).

This parallelogram is a fundamental region for the group

$$T' = \{\tau^n \mu^m \mid n, m \in \mathbb{Z}\} \cong \mathbb{Z} \times \mathbb{Z}$$

of translations, i.e., the images $\tau^n \mu^m(d)$ cover the whole plane.

Let $\alpha \in T$ be any translation. Then $\alpha(P)$ lies in a parallelogram $\tau^n \mu^m(d)$, i.e., $\mu^{-m}\tau^{-n}\alpha(P)$ lies in $d$. We want to show $\alpha \in T'$. Then we will have proven $T = T'$ and thus the assertion. But it is sufficient to show that $\beta = \mu^{-m}\tau^{-n}\alpha \in T'$.

If $\beta(P)$ is one of the vertices of $d$, then $\beta \in T'$, and the assertion is proven. $\beta(P)$ cannot lie on one of the edges $P, \tau(P)$ or $P, \mu(P)$, otherwise we would have a contradiction to the definition of $\tau$ or $\mu$. It also cannot lie on one of the other two edges, otherwise $\mu^{-1}\beta$ or $\tau^{-1}\beta$ would be a translation by a shorter distance than $\tau$ or $\mu$. Let $s$ be the distance from $P$ to $\mu(P)$. If $\beta(P)$ were in the open disc $E$ with center $P$ and radius $s$ (see Fig. 6.5), then $\beta(P)$ would be closer to $P$ than $\mu(P)$ in contradiction to the definition of $\mu$. It cannot also lie in the disc $E_1$ with the same radius around the point $\mu\tau(P)$, otherwise $\mu^{-1}\tau^{-1}\beta(P)$ would be closer to $P$ than $\mu(P)$. But the discs $E$ and $E_1$ cover all of $d$. Thus, $\beta(P)$ is one of the vertices of $d$, and the assertion is proven.                                                    □

The subgroup of translations of a decomposition of the plane into congruent squares is isomorphic to $\mathbb{Z} \times \mathbb{Z}$, as one can easily show.

**Exercises**

1. Show that a group $G$ of isometries of $\mathbb{R}^2$ is discontinuous, if and only if for every point $P \in \mathbb{R}^2$ and for every disc $D \subset \mathbb{R}^2$ the orbit of $P$ only has finitely many points in $D$.

2. Give a presentation of the subgroup of translations of $G_{(4,4)}$.

# Chapter 7
# Finite Groups

This chapter is about groups of finite order. In the first section, we present a class of examples of finite groups that will play a role later on.

By Lagrange's theorem, the orders of the subgroups of a finite group must be divisors of the order of the group. However, the theorem makes no statements about whether such subgroups actually exist. Such existence statements are given by the Sylow theorems presented in the second section. We then study some groups of small order with the help of the Sylow theorems. The last two sections deal with finite groups in Euclidean geometry and on the sphere.

## 7.1 An Example

In certain cases, a given presentation says a lot about the associated group. We want to illustrate this with an example:

**Example 7.1** The group $G_{m,n} = \langle x, y \mid x^m, y^n, xy = y^2x \rangle$ for $m, n \geq 2$ has finite order.

We can indeed transform every word $w \in G_{m,n}$ into a word from the following set:

$$\{y^k x^j \mid 0 \leq k < m, \ 0 \leq j < n\}. \tag{7.1}$$

The proof is similar to that of Theorem 5.5 on page 97: Every word $w$ in $x, y$ has the form $w = x^{\epsilon_1} y^{\delta_1} x^{\epsilon_2} y^{\delta_2} \ldots x^{\epsilon_k} y^{\delta_k}$. Because of the relations $x^m, y^n$ we can assume that $0 \leq \epsilon_i < m$ and $0 \leq \delta_i < n$. With the help of the relation $xy = y^2x$ we push each $y$ in $w$ to the left (at the cost of increasing the power), until all $y$ come before all $x$. Then we normalize both powers with the relations $x^m$ and $y^n$ and thus obtain a word in the set (7.1). It follows that $|G_{m,n}| \leq m \cdot n$.

© The Author(s), under exclusive license to Springer-Verlag GmbH, DE,
part of Springer Nature 2024
S. Rosebrock, *Visual Group Theory*, Springer Undergraduate Mathematics Series,
https://doi.org/10.1007/978-3-662-69365-0_7

Beware, the elements of (7.1) are not necessarily all different. For $n$ even, for example, if in $xyx^{-1} = y^2$ you take both sides to the power of $n/2$, you get $xy^{n/2}x^{-1} = y^n = 1$ and thus $y^{n/2} = 1$. If $n$ is a power of 2, you can iterate this trick (see Exercise 1) and get $y = 1$. In this case we have $G_{m,n} = \mathbb{Z}_m$.

On the other hand, GAP easily convinces us that $G_{3,7}$ has order $3 \cdot 7 = 21$:

```
gap> m:=3;;n:=7;;
gap> freigrp := FreeGroup("x","y");;
gap> x:=freigrp.1;;y:=freigrp.2;;
gap> gmn := freigrp/[x^m, y^n, x*y*x^-1*y^-2];;
gap> Order(gmn);
21
gap> e:=Elements(gmn);
[ <identity ...>, x, x^2, y, x*y, x^2*y, y*x, x^2*y*x,
  x*y*x, x^2*y^2, y^2, x*y^2, x*y*x*y, y*x*y, x^2*y*x*y,
  x*y^3, y^3, x^2*y^3, x*y*x*y*x, y*x*y*x, x^2*y*x*y*x ]
```

This group is not abelian, because $xy$ and $yx$ are different. The group is therefore different from $\mathbb{Z}_{21}$ and from $\mathbb{Z}_3 \times \mathbb{Z}_7$. So we have found a small group that has not yet appeared.

If we perform the same calculation for $n = 6$, we get from GAP the output

```
[ <identity ...>, x, x^2 ]
```

The groups $G_{m,n}$ have already been encountered in Sect. 6.3 as semidirect products. The conditions mentioned there translate to:

$$n \text{ must be odd and } 2^m \equiv 1 \bmod n.$$

The above case $m = 3$ and $n = 7$ is therefore a semidirect product of $\mathbb{Z}_3$ with $\mathbb{Z}_7$: $G_{3,7} = \mathbb{Z}_7 \rtimes \mathbb{Z}_3$. That $G_{3,7}$ has 21 elements follows from Exercise 1. of Sect. 6.3.

## Exercises

1. Show $G_{m,n} = \mathbb{Z}_m$, if $n$ is a power of 2.
2. Prove the output of GAP: $G_{3,6} = \mathbb{Z}_3$.
3. Prove that $G_{2,3} = D_3$.

## 7.2  The Sylow Theorems

Lagrange's theorem (see Theorem 3.14) states that the orders of subgroups of finite groups can only take certain values, they must be divisors of the group order. Whether such groups really exist isn't clear. The following *First Sylow Theorem* makes such a statement. This theorem was proven by Sylow in 1872.

**Theorem 7.2 (First Sylow Theorem)** *Let G be a group of order $p^l m$, where p does not divide m and $l \geq 1$. Then G contains a subgroup of order $p^l$.*

Such a subgroup of order $p^l$ of $G$ is called a *p-Sylow subgroup* or *Sylow subgroup*. Before we prove Theorem 7.2, we need some preparation:

Let $X$ be a $G$-set. Then $G$ operates on the $k$-element subsets of $X$. If $U \subset X$ is a $k$-element subset, then for $g \in G$ the subset $gU = \{gu \mid u \in U\}$ also has $k$ elements, because each group element $g \in G$ permutes the elements of $X$ (see page 65).

The number of $k$-element subsets of an $n$-element set is for $1 \leq k \leq n$ given by the *binomial coefficient* $\binom{n}{k}$ and is calculated by:

$$\binom{n}{k} = \frac{n!}{(n-k)!k!} = \frac{n(n-1)(n-2)\cdots(n-k+1)}{k(k-1)(k-2)\cdots 1}.$$

The exclamation mark denotes the *factorial* of a number, i.e.

$$n! = n(n-1)(n-2)(n-3)\cdots 2 \cdot 1.$$

In GAP, binomial coefficients can be calculated directly:

```
gap> Binomial(6,3);
20
```

**Lemma 7.3** *Let $n = p^l m$, $l \geq 1$, where p does not divide m. Then $\binom{n}{p^l}$ is not divisible by p.*

**Proof** We have

$$\binom{n}{p^l} = \frac{n(n-1)(n-2)\cdots(n-j)\cdots(n-p^l+1)}{p^l(p^l-1)(p^l-2)\cdots(p^l-j)\cdots 1}.$$

We assign to each factor $(n-j)$ the factor $(p^l-j)$ in the denominator. If we can prove that $p$ divides $(n-j)$ as often as $(p^l-j)$, then we are done, because no $p$ remains in the numerator which may not be canceled by a $p$ in the denominator. We decompose $j$ into $j = p^e k$, where $k$ is not divisible by $p$. It follows that $e < l$, because the denominator factor $(p^l - j)$ is greater than 0. Since $n$ and $p^l$ are divisible by $p^e$, so are $(n-j)$ and $(p^l-j)$. But they are not divisible by $p^{e+1}$.  □

**Lemma 7.4** *Assume the finite group G acts on the set of all subsets of G by left multiplication. Let U be a subset of G. Then the order of the stabilizer $G(U)$ is a divisor of $|U|$.*

**Proof** $U$ consists of the orbits $G(U)g$, where $g \in U$. These orbits are right cosets according to Example 4.21. Thus, $U$ consists of a union of right cosets. The number of elements of $U$ is therefore a multiple of $|G(U)|$.  □

We now prove the first Sylow theorem:

**Proof** Let $X$ be the set of all subsets of $G$ with exactly $p^l$ elements. $G$ operates on $X$ through left multiplication, because each $p^l$-element subset of $G$, via multiplication by a group element, transforms into another $p^l$-element subset, as explained above.

We will show that one of these subsets has a stabilizer of order $p^l$. Since stabilizers are subgroups, this proves the claim.

Let $|G| = n = p^l m$. It holds that $|X| = \binom{n}{p^l}$, and according to Lemma 7.3 $|X|$ is not divisible by $p$. With respect to the operation of $G$ on $X$ we decompose $X$ according to Theorem 4.13 into orbits:

$$|X| = |B_1| + |B_2| + \cdots + |B_r|.$$

Since $p$ does not divide $|X|$, there is an orbit $B_i$, the number of elements of which is not divisible by $p$. Let $U \in X$ be a subset that belongs to the orbit $B_i$. Because $U \in X$, $p^l$ is the number of elements of $U$. According to Lemma 7.4 the number of elements of $|G(U)|$ is therefore a $p$-power.

The orbit-stabilizer theorem (Theorem 4.35) yields here:

$$p^l m = |G| = |G(U)| \cdot |B_i|.$$

Since $p$ does not divide $|B_i|$, it follows that $|G(U)| = p^l$, and $G(U)$ is the subgroup in question. □

In GAP we can easily obtain Sylow subgroups.

```
gap> G:=SymmetricGroup(5);;
gap> SylowSubgroup(G,2);
Group([ (1,2), (3,4), (1,3)(2,4) ])
gap> Order(last);
8
```

The following *Second Sylow Theorem* provides more precise information about Sylow subgroups:

**Theorem 7.5 (Second Sylow Theorem)** *Let $G$ be a finite group and $J < G$. Let $p$ be a prime number which divides $|J|$. Let $H$ be a $p$-Sylow subgroup of $G$. Then there is a subgroup $H' = gHg^{-1}$, conjugate to $H$, such that $J \cap H'$ is a Sylow subgroup of $J$.*

**Proof** $G$ acts by left multiplication on the set $X = G/H$ of left cosets of $H$: $g \cdot (g'H) = (gg')H$. This operation is transitive, because by multiplication by a suitable $g \in G$ any given coset can be transformed into any other. The stabilizer $G(x)$ of the element $x = 1H$ is $H$ itself. The stabilizer of $gx$ for any $g \in G$ is the conjugate subgroup $gHg^{-1}$, because

$$gHg^{-1} \cdot gx = gHg^{-1} \cdot gH = gH \cdot H = gH = gx.$$

The number of elements of $X$ is given by $|G|/|H|$, and because $H$ is a $p$-Sylow subgroup, it is not divisible by $p$.

We only let the elements of $J$ act on $X$ and decompose $X$ into disjoint orbits under this operation. Because $|X|$ is coprime to $p$, there is an orbit $B$ whose length is not divisible by $p$. Let $gx \in B$. As mentioned above, $H' = gHg^{-1}$ is then the stabilizer of the element $gx$ with respect to the operation of $G$. If we restrict the operation to $J$, only the elements of $H' \cap J$ stabilize. By Theorem 4.34 the number of cosets equals the length of the orbit, so we have $[J : H' \cap J] = |B|$, and therefore this index is coprime to $p$. A conjugate subgroup always has the same number of elements as the subgroup (see Theorem 4.29), and because $J$ is assumed to be a $p$-group, $H' \cap J$ is also a $p$-group (the order of a subgroup has to divide the order of the group). Because its index in $J$ is coprime to $p$, it is a $p$-Sylow subgroup of $J$.                                                                                          □

**Corollary 7.6** *Let $G$ be a finite group. The following statements hold:*

1. *Let the subgroup $J < G$ be a $p$-group. Then $J$ is contained in a $p$-Sylow subgroup.*
2. *All $p$-Sylow subgroups of $G$ are conjugate to each other.*

**Proof** A $p$-group $J$ has a $p$-power as order and thus only one Sylow subgroup, namely itself. From the Second Sylow Theorem it follows that to a $p$-Sylow subgroup $H < G$ there corresponds a conjugate subgroup $H' < G$ with $J \cap H' = J$. The last condition just means $J \subset H'$. Item 1 now follows from the fact that, because $H'$ has as many elements as $H$, $H'$ is a $p$-Sylow subgroup.

For 2, we assume that $J$ and $H$ are $p$-Sylow subgroups of $G$. In the proof of 1 we have seen that there is a conjugate $p$-Sylow subgroup $H'$ to $H$, with $J \subset H'$. The subgroups $J$ and $H'$ have the same order, so $J = H'$, and $J$ is conjugate to $H$.      □

We let a finite group $G$ operate on its $s$-element subsets by conjugation. If $H$ is an $s$-element subgroup, then the orbit consists of the subgroups conjugate to $H$:

$$G_H = \{gHg^{-1} \mid g \in G\}.$$

$H$ is a normal subgroup if and only if $gHg^{-1} = H$ for all $g \in G$, i.e. if $G_H$ contains only $H$ itself. The stabilizer with respect to the operation of conjugation of $s$-element subsets is called the *normalizer*:

$$G(H) = \{g \in G \mid gHg^{-1} = H\}.$$

$G(H) = G$ if $H$ is a normal subgroup in $G$. The further away $H$ is from being a normal subgroup, the smaller the normalizer.

If $g \in G$ has the property $gHg^{-1} = H$, then the same holds for every element of the coset $gH$. This is due to Lemma 3.13: If $j \in gH$ and $gH = Hg$, then $gH = jH$ and $Hg = Hj$ and thus $jH = Hj$. The normalizer $G(H)$ thus consists of whole cosets.

**Theorem 7.7** *If $H < G$, then $G(H) < G$ and $H \lhd G(H)$.*

**Proof** To show the closure, we use Lemma 3.38: If $gHg^{-1} \in H$, then we can replace $H$ by $gHg^{-1}$ in $jHj^{-1}$. Because $jHj^{-1} \in H$, it follows that

$jgHg^{-1}j^{-1} \in H$. The identity element of $G$ is in $G(H)$. If $g \in G(H)$, then $g^{-1} \in G(H)$. From $gH = Hg$ it follows that $g^{-1}H = Hg^{-1}$ by multiplying by $g^{-1}$ from the left and from the right.

We now show $H \lhd G(H)$: $H$ is a subgroup of $G(H)$, because every $h \in H$ satisfies $hHh^{-1} = H$. The conjugation permutes the elements of $H$. This implies also that $H$ is a normal subgroup in $G(H)$ (but not necessarily in $G$).  □

In GAP we consider $S_5(S_3)$, where the subgroup $S_3$ permutes the elements 1,2,3:

```
gap> G:=SymmetricGroup(5);;
gap> Normalizer(G,Subgroup(G,[(1,2,3)]));
Group([ (1,2,3), (4,5), (2,3) ])
gap> Elements(last);
[ (), (4,5), (2,3), (2,3)(4,5), (1,2), (1,2)(4,5), (1,2,3),
  (1,2,3)(4,5), (1,3,2), (1,3,2)(4,5), (1,3), (1,3)(4,5) ]
```

It is clear that in this normalizer only those elements which do not permute 4 or 5 with 1, 2 or 3 may occur.

Let $k$ be the number of subgroups conjugate to $H$. Then the orbit-stabilizer theorem gives

$$|G| = |G(H)| \cdot k.$$

So the number of subgroups conjugate to $H$ is equal to the index $[G : G(H)]$. If $H$ is a $p$-Sylow subgroup of $G$, then by Corollary 7.6 all groups of the same order are conjugate to $H$. The number of $p$-Sylow subgroups is then $k = [G : G(H)]$.

**Theorem 7.8 (Third Sylow Theorem)** *Let $G$ be a group of order $p^l m$, where $p$ does not divide $m$ and $l \geq 1$. Let $k$ be the number of $p$-Sylow subgroups of $G$. Then $p$ is a divisor of $k - 1$ and $k$ is a divisor of $m$.*

**Proof** As just determined, for the number of $p$-Sylow subgroups $k = [G : G(H)]$ given a $p$-Sylow subgroup $H < G$ with respect to the operation of conjugation. Because $H \subset G(H)$, $k = [G : G(H)]$ divides $m = [G : H]$.

To show that $p$ divides $k - 1$, we consider the set $M = \{H = H_1, H_2, \ldots, H_k\}$ of $p$-Sylow subgroups. We now consider conjugation only with elements of $H$ for an arbitrary $p$-Sylow subgroup $H_j \in M$. We decompose $M$ with respect to this operation into orbits $B_1, \ldots, B_r$. So, for example, $B_i = \{hH_jh^{-1} \mid h \in H\}$. Let $B_1$ be the orbit of $H$.

Let $N_j = G(H_j) = \{g \in G \mid gH_jg^{-1} = H_j\}$ be the normalizer of $H_j$. The orbit $B_i$ consists of only one subgroup $H_j$ if and only if $H \in N_j$. $H \in N_j$ means precisely that $hH_jh^{-1} = H_j$ for all $h \in H$. If $H \in N_j$, then $H$ and $H_j$ are $p$-Sylow subgroups of $N_j$, because they are $p$-Sylow subgroups of $G$. Then $H$ and $H_j$ are conjugate by Corollary 7.6. $H_j$ is normal in $N_j$, and therefore $H = H_j$. So there is only one $H$-orbit of length 1, namely $B_1 = \{H\}$. The length of the other $H$-orbits $B_2, \ldots, B_r$ are $p$-powers, because of the orbit-stabilizer theorem

$$p^l = |H| = |H(H_t)| \cdot |B_i|, \qquad r \geq i \geq 2,$$

if $B_i$ is the orbit of $H_t$. It follows from the class equation (Theorem 4.38) that $k = 1 + p^{\epsilon_2} + \cdots + p^{\epsilon_r}$ and thus $p$ divides $k - 1$. □

## Exercises

1. Calculate the normalizer $G(G)$ for any group $G$.
2. Let $H = \langle 3 \rangle$ be a subgroup of the finite cyclic group $\mathbb{Z}_{15}$. Calculate the normalizer $\mathbb{Z}_{15}(H)$.
3. Consider $D_4$, and a reflection $s \in D_4$. Let $H = \langle s \rangle$. Calculate the normalizer $D_4(H)$.

## 7.3 Some Groups of Small Order

Cauchy proved the following theorem as early as 1846 for permutation groups:

**Theorem 7.9 (Cauchy's Theorem)** *Let $G$ be a finite group and $p$ a prime number that divides the order of the group. Then $G$ contains an element of order $p$.*

*Proof* Let $H < G$ be a Sylow subgroup of order $p^l$ and $1 \neq h \in H$. The subgroup generated by $h$ divides the group order, so the order of $h$ divides the order of $H$. Thus, $|h| = p^k$, $0 < k \leq l$. Then $g = h^{p^{k-1}}$ has order $p$, because $g^p = h^{p \cdot p^{k-1}} = h^{p^k} = 1$. The order of $g$ cannot be smaller than $p$ because $g \neq 1$, and it must divide the order of $H$. □

The groups of orders up to five are already known to us: The trivial group is the only group of order 1. The groups of prime order are all cyclic (Corollary 3.20), and otherwise there is only the Klein four-group $\mathbb{Z}_2 \times \mathbb{Z}_2$ (see Definition 2.20 and Exercise 9. of Sect. 3.1) with 4 elements.

**Theorem 7.10** *Up to isomorphism, there are exactly two groups of order six, the cyclic group $\mathbb{Z}_6$ and the Dihedral group $D_3$.*

*Proof* Due to Theorem 7.9, every group of order six must contain an element $x$ of order 3 and an element $y$ of order 2. The six elements

$$G = \{x^i y^j \mid 0 \leq i \leq 2, 0 \leq j \leq 1\}$$

are all different, because an equation $x^i y^j = x^m y^n$ could be written as $x^{i-m} = y^{n-j}$, but since every power of $x$, except the identity element, has order 3 and every power of $y$, except the identity element, has order 2, it follows that $i = m$ and $n = j$. The set $G$ is therefore already the whole group. Since $yx \neq 1, x, x^2, y$, it follows that $yx = xy$ or $yx = x^2y$. The two possibilities for groups of order 6 are therefore:

$$P_1 = \langle x, y \mid x^3, y^2, yx = xy \rangle \text{ and } P_2 = \langle x, y \mid x^3, y^2, yx = x^2y \rangle.$$

No more relations are necessary, as we can determine the product of any two elements of $G$ using the relations. Since we already know two groups of order 6, these must be the two.                                                                     □

The presentation $P_1$ has already appeared as a presentation of the direct product $\mathbb{Z}_3 \times \mathbb{Z}_2$ (see Corollary 6.6 on page 116). In Exercise 4. of Sect. 6.1 it was proven that $\mathbb{Z}_3 \times \mathbb{Z}_2 = \mathbb{Z}_6$.

The presentation $P_2$ is the group $G_{2,3}$ of Sect. 7.1. The relation $yx = x^2y$ in $P_2$ can be rewritten as $yxyx$, and we get

$$P_2' = \langle x, y \mid x^3, y^2, yxyx \rangle.$$

This presentation has already appeared as a presentation of the group $D_3$ in Theorem 5.5.

Seven is a prime number, so there is only the group $\mathbb{Z}_7$ of order seven.

Quite analogously to the proof of Theorem 7.10 one shows:

**Theorem 7.11** *If $p$ is a prime number with $p \geq 3$ and $G$ is a group of order $2p$, then $G$ is isomorphic to $\mathbb{Z}_{2p}$ or $D_p$.*

One can prove that there are 5 groups of order 8 (see for example [Cig95]). The groups $\mathbb{Z}_8, \mathbb{Z}_4 \times \mathbb{Z}_2, \mathbb{Z}_2 \times \mathbb{Z}_2 \times \mathbb{Z}_2$ obviously all have order 8 and are different. There is only one element of order 8 in the first group, and the second group contains an element of order 4, unlike the third group. The group $D_4$ is non-commutative and thus different from the above three. In Exercise 6. of Sect. 5.1 we have already got to know the quaternion group. It is the last group of order 8. There we also proved that it is not isomorphic to any of the above groups.

To analyze the groups of order 9, we prove:

**Theorem 7.12** *If $p$ is a prime number, then all groups of order $p^2$ are abelian.*

**Proof** Let $G$ be a group of order $p^2$, and $p$ be prime. According to Theorem 4.44 on page 80, the center $C(G)$ of $G$ is nontrivial. It has $p$ or $p^2$ elements, because it is a subgroup of $G$. The factor group $G/C(G)$ therefore has order $p$ or is trivial. In the latter case, $G = C(G)$ and therefore $G$ is abelian.

Let $G/C(G)$ have order $p$. Let $x \in G$ not be in the center. Then there are more elements of $G$ commuting with $x$ than commuting with all $g \in G$. This means that the centralizer $Z(x) = \{g \in G \mid gx = xg\}$ contains more elements than the center. Because $Z(x)$ is a subgroup of $G$, it follows that $|Z(x)| = p^2$ and thus $G = Z(x)$. However, this implies that $x$ must be in the center, contradicting our assumption. Therefore, the case $|G/C(G)| = p$ does not occur, and $G$ is abelian.                                   □

We already know two groups of order 9:

$$\mathbb{Z}_9 = \langle x \mid x^9 \rangle \quad \text{and} \quad \mathbb{Z}_3 \times \mathbb{Z}_3 = \langle x, y \mid x^3, y^3, yx = xy \rangle.$$

These groups are different because the latter, unlike the first, does not contain an element of order 9. In fact, these are the only two groups of order 9. If a group of order 9 contains an element of order 9, it is the group $\mathbb{Z}_9$. Otherwise, all nontrivial elements have order 3. Let $x$, $y$ be two such elements, where $x$ is not a power of $y$. Since the group is abelian according to Theorem 7.12, these two elements must commute. We get the presentation $\langle x, y \mid x^3, y^3, yx = xy \rangle$.

```
gap> SmallGroupsInformation(9);
```

```
There are 2 groups of order 9.
[...]
```

In GAP we conveniently get all groups of a given small order. The group $G$ constructed here is $\mathbb{Z}_3 \times \mathbb{Z}_3$.

```
gap> SmallGroup(9,1);
<pc group of size 9 with 2 generators>
gap> G:=SmallGroup(9,2);
<pc group of size 9 with 2 generators>
gap> Elements(G);
[ <identity> of ..., f1, f2, f1^2,
f1*f2, f2^2, f1^2*f2, f1*f2^2,
f1^2*f2^2 ]
```

According to Theorem 7.11 there are only two groups of order 10, the groups $D_5$ and $Z_{10}$ and also only the groups $D_7$ and $Z_{14}$ of order 14.

Here are the groups of order 12 in GAP:

```
gap> A:=AllGroups(12);;
gap> List(A, cs -> StructureDescription(cs));
[ "C3 : C4", "C12", "A4", "D12", "C6 x C2" ]
```

So there are the groups $\mathbb{Z}_{12}$, $\mathbb{Z}_2 \times \mathbb{Z}_6$, $D_6$, $A_4$ and the semidirect product $\mathbb{Z}_3 \rtimes \mathbb{Z}_4$. This last group is the group $G_{4,3}$ of Sect. 7.1 (see also Exercise 2).

We now know all isomorphism types of groups up to order 15 (according to Corollary 7.14 there is only one group of order 15):

| Order | Number | Abelian | Non-Abelian |
|-------|--------|---------|-------------|
| 1 | 1 | Trivial group $\{e\}$ | |
| 2 | 1 | $\mathbb{Z}_2$ | |
| 3 | 1 | $\mathbb{Z}_3 \cong A_3$ | |
| 4 | 2 | $\mathbb{Z}_4, D_2 \cong \mathbb{Z}_2 \times \mathbb{Z}_2$ | |
| v5 | 1 | $\mathbb{Z}_5$ | |
| 6 | 2 | $\mathbb{Z}_6 \cong \mathbb{Z}_2 \times \mathbb{Z}_3$ | $D_3 \cong S_3$ |
| 7 | 1 | $\mathbb{Z}_7$ | |
| 8 | 5 | $\mathbb{Z}_8, \mathbb{Z}_4 \times \mathbb{Z}_2, \mathbb{Z}_2 \times \mathbb{Z}_2 \times \mathbb{Z}_2$ | $D_4, Q$ |
| 9 | 2 | $\mathbb{Z}_9, \mathbb{Z}_3 \times \mathbb{Z}_3$ | |
| 10 | 2 | $\mathbb{Z}_{10} \cong \mathbb{Z}_2 \times \mathbb{Z}_5$ | $D_5$ |
| 11 | 1 | $\mathbb{Z}_{11}$ | |
| 12 | 5 | $\mathbb{Z}_{12}, \mathbb{Z}_6 \times \mathbb{Z}_2$ | $D_6, A_4, \mathbb{Z}_3 \rtimes \mathbb{Z}_4$ |
| 13 | 1 | $\mathbb{Z}_{13}$ | |
| 14 | 2 | $\mathbb{Z}_{14} \cong \mathbb{Z}_2 \times \mathbb{Z}_7$ | $D_7$ |
| 15 | 1 | $\mathbb{Z}_{15}$ | |

In [Joy02], a table of all groups up to order 25 is given, including presentations.

The number of groups with $n$ elements increases significantly with increasing $n$. For example, there are exactly 49,487,365,422 pairwise non-isomorphic groups with 1024 elements.

**Theorem 7.13** *Let $p > q$ be prime numbers, and let $G$ be a group of order $p \cdot q$. Assume that $q$ does not divide $p - 1$. Then $G$ is isomorphic to $\mathbb{Z}_{p \cdot q}$.*

**Proof** Let $G$ be any group of order $p \cdot q$. According to the Third Sylow Theorem, the number $k$ of $p$-Sylow subgroups of $G$ is a divisor of $q$, and $p$ divides $k - 1$. Because $p > q$, it follows that there is only one $p$-Sylow subgroup $H < G$. Any subgroup conjugate to $H$ has the same number of elements, so it must be equal to $H$, and therefore $H$ is normal in $G$.

Similarly, it can be seen that there is only one $q$-Sylow subgroup $H'$, which is also normal in $G$. The number $m$ of $q$-Sylow subgroups of $G$ must be a divisor of $p$ (so $m = 1$ or $m = p$), and $q$ must divide $m - 1$. We assumed $q$ does not divide $p - 1$, hence $m = 1$.

The order of $H \cap H'$ divides $p$ and $q$, so it follows that $H \cap H' = \{1\}$. Since $H \cdot H'$ is a subgroup with more than $p$ elements, it follows that $G = H \cdot H'$. The characterization of the direct product, Theorem 6.3, shows that $G = H \times H'$. The claim now follows from Exercise 4. of Sect. 6.1.                                                         □

**Corollary 7.14** *Up to isomorphism, the group $\mathbb{Z}_{15}$ is the only group of order 15.*

We already know two groups of order 21, the group $\mathbb{Z}_{21}$ and the semidirect product $G_{3,7} = \mathbb{Z}_7 \rtimes \mathbb{Z}_3$ of Sect. 7.1. The Third Sylow Theorem does not rule out that there can be seven conjugate 3-Sylow subgroups. This happens in the group $G_{3,7}$, and we cannot argue as in Theorem 7.13.

# Exercises

1. Prove that if $p$ is prime, then there are exactly two groups of order $p^2$, the groups $\mathbb{Z}_{p^2}$ and $\mathbb{Z}_p \times \mathbb{Z}_p$, and these are not isomorphic.
2. Is the alternating group $A_4$ isomorphic to the dihedral group $D_6$? If you get stuck, you can use GAP to view the orders of the elements. For the group $A_4$, this can be done with
   `A4:=AlternatingGroup(4);; List(Elements(A4), i->Order(i));`
3. Show that the group $\mathbb{Z}_{77}$ is the only group of order 77.
4. Prove that the group $A_5$ does not contain subgroups of orders 15 and 30.
5. How many elements of order 4 does the group $S_5$ contain?
6. Consider the group $S_n$, the group of permutations of the set $T_n = \{1, 2, 3, \ldots, n\}$. The *support* of an element is the set of numbers in $T_n$ that appear in the permutation notation of the element, excluding the cycles of length 1. For example, the element $(1, 5, 6)(3, 9)$ has support $\{1, 3, 5, 6, 9\}$. Show:

(a) 2 elements of $S_n$ commute if their supports are disjoint.

(b) For $p, q \in S_n$, $pq$ and $qp$ have the same cycle structure.

(c) With which element must one conjugate to generate $(1, 2, 9)(5, 3)$ from $(1, 5, 6)(3, 9)$?

(d) Determine the class equation of the group $S_4$.

## 7.4  The Orthogonal Group

In Definition 3.5, the group $O_2$ is defined as the subgroup of the symmetry group of the plane that leaves the origin fixed. According to Theorem 1.6, it consists of reflections and rotations only.

**Theorem 7.15** *Let $G < O_2$ be a finite group. Then $G \cong D_n$ or $G \cong D_n^+ \cong \mathbb{Z}_n$ for some $n \in \mathbb{N}$.*

**Proof** If all elements of $G$ are rotations, it suffices to show that $G$ is cyclic. By Theorem 2.15, it then follows that $G = D_n^+ = \mathbb{Z}_n$. The proof is very similar to that of Theorem 3.10: If $G$ is the trivial group, then $G = \mathbb{Z}_1$. Otherwise, $G$ contains a nontrivial rotation $d_\alpha$ by an angle $0 < \alpha < 2\pi$, where $\alpha$ is the smallest angle that occurs in such a rotation.

We prove $G = \langle d_\alpha \rangle$. If $d_\beta \in G$, then $\beta = t \cdot \alpha + \psi$ for some $t \in \mathbb{N} \cup \{0\}$ with $0 \leq \psi < \alpha$. Since $d_\psi = d_\beta d_{t\alpha}^{-1}$ is also in $G$, $\psi = 0$ must hold, because $0 \leq \psi < \alpha$ and $\alpha$ was chosen to be minimal. It follows that $d_\beta = d_{t\alpha} = d_\alpha^t$, and the claim is proven for the case where $G$ contains rotations only.

In the case where $G$ contains at least one reflection $s$, we consider the subgroup of all rotations, which is $D_n^+ < G$, as shown above. The elements

$$E = \{id, d_\alpha, d_\alpha^2, \ldots, d_\alpha^{n-1}, s, sd_\alpha, sd_\alpha^2, \ldots, sd_\alpha^{n-1}\}$$

are therefore in $G$, and these are according to Sect. 3.2 exactly the elements of the group $D_n$. Therefore, $E = D_n < G$. If $G$ contains additional elements, they must be reflections. Let $s' < G$ be any reflection. The product of two reflections is a rotation according to Theorem 1.9, so $s \cdot s' = d_\gamma \in G$ for a rotation $d_\gamma$. Since $s, d_\gamma \in E$ it follows that $s' \in E$, and therefore $G = D_n$.  $\square$

**Theorem 7.16** *Let $G < \mathcal{E}$ be discontinuous. Then the stabilizer $G(P)$ of a point $P \in \mathbb{R}^2$ is finite, and either a cyclic group or a dihedral group.*

**Proof** Any circle $K$ with center $P$ is mapped onto itself by $G(P)$. Let $Q$ be a point on $K$ and $D$ a disc that contains all of $K$. Since $G$ is discontinuous, the orbit of the point $Q$ under $G(P) < G$ can only contain finitely many points, and therefore $G(P)$ is finite. According to Theorem 7.15, $G(P)$ is cyclic or dihedral.  $\square$

**Corollary 7.17** *Let $G < \mathcal{E}$ be discontinuous, and its translation subgroup $T = G \cap \mathcal{T}$ be trivial. Then $G$ is finite, and either a cyclic group or a dihedral group.*

**Proof** If $G$ contains no translations, then it fixes a point $P$ (see Exercise 1). So $G$ is the stabilizer of a point, and we apply Theorem 7.16.                                             □

If $G < \mathcal{E}$ is discontinuous, then according to Theorem 6.21 the translation subgroup $T < G$ may be trivial, and this case is covered by Corollary 7.17. If $T$ is infinite cyclic (so isomorphic to $\mathbb{Z}$) then $G$ is a frieze group, according to the definition of friezes. There are 7 different types of groups of friezes, where not only the isomorphism of the groups is considered, but also which types of isometries belong to the group (see for example [Hen12]).

In the remaining case, $T \cong \mathbb{Z} \times \mathbb{Z}$, $G$ is called a *crystallographic group*. It can be shown that there are 17 types of planar crystallographic groups. A detailed treatment of the crystallographic groups and friezes can be found in [Lyn85] and also in [Hen12]. We only want to prove here the so-called *crystallographic restriction*:

**Theorem 7.18** *Let $G < \mathcal{E}$ be a finite subgroup of a crystallographic group. Then every rotation of $G$ has order $1, 2, 3, 4$ or $6$ and $G$ is isomorphic to one of the groups $\mathbb{Z}_n$ or $D_n$ for $n = 1, 2, 3, 4$ or $6$.*

**Proof** The second statement follows from the first by Theorem 7.15. To prove the first statement, we assume that $G$ contains a rotation $d$ of order $n \geq 2$ about the point $P$, and $n$ is chosen maximally. Let $\tau \in G$ be a translation of shortest length and $\tau(P) = Q$.

Then the $n$ points

$$Q, d(Q), d^2(Q), \ldots, d^{n-1}(Q)$$

have the same distance from their respective neighbor on the circle with radius $|\tau| = |PQ|$ around $P$. For $n = 6$ this distance is exactly $|\tau|$, for $n > 6$ it is smaller than $|\tau|$. So the distance from $\tau(P)$ to $d\tau(P)$ in that case is smaller than $|\tau|$ (see Fig. 7.1).

We have

$$d\tau d^{-1}\tau^{-1}(\tau(P)) = d\tau d^{-1}(P) = d\tau(P).$$

The translation $\alpha = d\tau d^{-1}\tau^{-1}$ thus maps $\tau(P)$ to $d\tau(P)$ and therefore, when $n > 6$, contradicts the minimality of $\tau$, having a shorter length than $\tau$.

**Fig. 7.1** The case $n > 6$

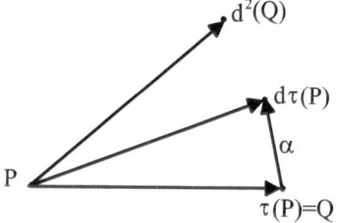

**Fig. 7.2**  The case $n = 5$                            $\bullet d(Q)$

                         $\bullet d^2(Q)$

          $\bullet \tau^{-1}(P)$         $\bullet P$            $\bullet Q = \tau(P)$

          $\bullet d^3(Q)$                   $\bullet d^2\tau^{-1}(P)$

                         $\bullet d^4(Q)$

If $n = 5$, the distance from $d^2\tau^{-1}(P)$ to $\tau(P)$ is smaller than $|\tau|$ (see Fig. 7.2). We have

$$d^2\tau^{-1}d^{-2}\tau^{-1}(\tau(P)) = d^2\tau^{-1}(P).$$

The translation $d^2\tau^{-1}d^{-2}\tau^{-1}$ thus maps $\tau(P)$ to $d^2\tau^{-1}(P)$ and therefore contradicts the minimality of $\tau$, having a shorter length than $\tau$.       □

## Exercises

1. Prove that if a group $G < \mathcal{E}$ contains no translations (except the trivial one), then there is a point $P$ of the plane that remains fixed under every isometry of $G$.
2. Prove that the group $O_2$ is generated by a reflection and all rotations.

## 7.5   Regular Decompositions of the 2-Sphere

In this section, we consider isometries of $\mathbb{R}^3$, specifically those that leave the origin fixed. The associated group is the orthogonal group $O_3$, as we have already established. This group operates on the sphere $S$. The *sphere* $S$ is the set of all points of distance 1 from the origin in $\mathbb{R}^3$.

What are typical elements of the group $O_3$, and how do they affect $S$? Let's consider a plane $E$ in $\mathbb{R}^3$ that passes through the origin. $E \cap S$ is a so-called *great circle* $g$ on the sphere (see Fig. 7.3). Reflection through $E$ maps the two halves of the sphere onto each other. Considering isometries of $S$, we can interpret this reflection as a reflection over the great circle $g$. Reflections can therefore be carried out over great circles on the sphere. Great circles are *straight lines* on the sphere. They realize shortest connections. If you fly from Frankfurt to New York, you do not fly on a latitude circle, but keep to the north. You come very close to Greenland.

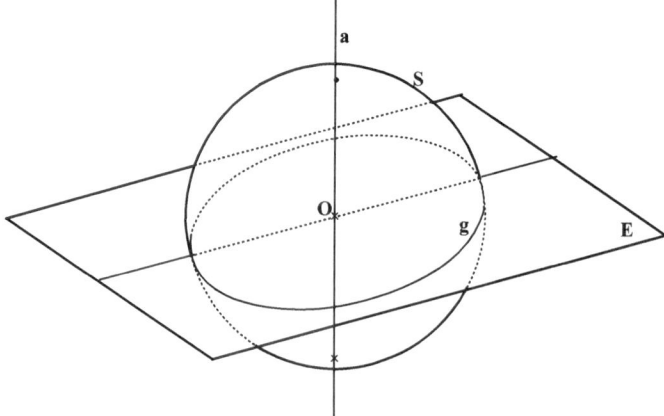

**Fig. 7.3**  Isometries on the sphere surface

Let's consider a line $a$ in $\mathbb{R}^3$ that passes through the origin. $a \cap S$ consists of 2 opposite points on the sphere, so-called *diametric points*. The sphere can be rotated about $a$ to obtain an isometry of $S$. Each pair of diametrical points on $S$ can therefore be used as a rotation center, and the associated isometry is a *spherical rotation*.

Any two different lines (great circles) $g$, $g'$ on $S$ have 2 intersection points. There are no parallel lines on $S$. The product of the associated reflections $s_g \circ s_{g'}$ results, according to Theorem 1.9, in a rotation by the double angle between $g$ and $g'$. The proof of Theorem 1.9 applies in a similar sense also to the sphere and not only to the plane. Translations cannot be performed on the sphere. It can be proven that every orientation-preserving isometry of the 2-sphere is a rotation (see Theorem 4.19 on page 69).

Since straight lines on the sphere are now defined, we can define triangles and more generally $n$-gons on the sphere. It is easy to see that the sum of the angles in a triangle on the sphere is always greater than $\pi$.

Next, we look for decompositions of the 2-sphere $S$ *of type (n,m)*. So we are looking for decompositions of $S$ into regular $n$-gons, arranged in such a way that always $m$ meet at one point. The $n$-gons may only intersect at their boundary edges and must not overlap. Such decompositions are called *regular decompositions*.

Here is an alternative definition of a regular decomposition. Let $T$ be a decomposition of the sphere or the Euclidean plane. A *flag* $(P, k, \sigma)$ in such a decomposition is a vertex $P \in T$ with adjacent edge $k \in T$ with adjacent $n$-gon $\sigma \in T$. For example, we have a flag in the cube in Fig. 1.8, if we consider the vertex 4 together with the edge $(4, 8)$ together with the quadrilateral $(4, 8, 7, 3)$. A decomposition is called *regular* if it is *flag-transitive*, i.e., if for any two flags there is an isometry of the decomposition that maps one flag onto the other. If $G$ is the symmetry group of $T$, then $G$ operates on the vertices and edges and $n$-gons. If the operation is flag-transitive, then it is also transitive on the vertices, on the edges and on the $n$-gons.

The regular decompositions of the Euclidean plane are of type $(3, 6)$, $(4, 4)$ and $(6, 3)$. We have dealt with the decomposition of type $(3, 6)$ in detail in Sect. 4.6, and the other two types appeared in the exercises of Sect. 4.6. See also Exercise 1. of Sect. 6.1.

**Theorem 7.19** *The regular decompositions of the 2-sphere are of type* $(n, 2)$, $(2, m)$, $(3, 3)$, $(4, 3)$, $(3, 4)$, $(3, 5)$ *and* $(5, 3)$, *for arbitrary* $n, m \in \mathbb{N}$.

*Proof* As one can easily show with elementary geometry, the sum of the angles in a regular $n$-gon on $S$ is greater than in the Euclidean plane, thus greater than $(n - 2)\pi$. Each individual angle is therefore greater than $(n - 2)\pi/n$. $m$ such angles sum up at a vertex to $2\pi$. So we have

$$m\frac{(n - 2)\pi}{n} < 2\pi.$$

This inequality is equivalent to

$$\frac{1}{m} + \frac{1}{n} > \frac{1}{2}.$$

Now it is easy to show that there can be no more than the above possibilities for the cell decompositions of $S$.                                                                     □

That the above types really exist can be seen by explicitly specifying the corresponding cell decomposition. We will do this now.

**The Decomposition of Type** $(3, 3)$   The sphere $S$ is to be decomposed into regular triangles, of which 3 should meet at each vertex. We obtain such a decomposition by embedding a tetrahedron in $\mathbb{R}^3$ in such a way that its vertices lie on $S$. Then we project the edges of the tetrahedron from the origin onto $S$ (see Fig. 7.4).

The symmetry group $G_{(3,3)}$ of this decomposition of $S$ is therefore isomorphic to the group of the tetrahedron. It follows that $G_{(3,3)} = S_4$, and this group has 24 elements, as we have already seen. If we look for a fundamental domain after

**Fig. 7.4** Decomposition of the 2-sphere into equilateral triangles

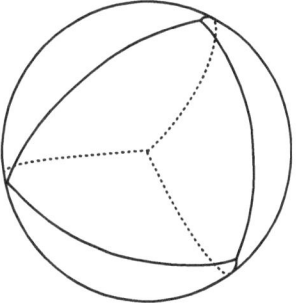

**Fig. 7.5**   Tetrahedron with
fundamental domain

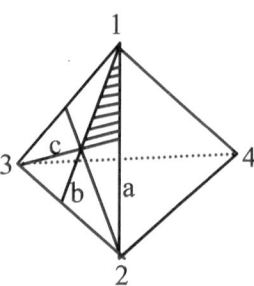

drawing all reflections of $G_{(3,3)}$ in $S$, we find, according to Theorem 4.57, a
generating system for $G_{(3,3)}$. From Fig. 7.5, we derive the following input for GAP.

```
gap> a:=(3,4);; b:=(2,3);; c:=(1,2);;
gap> Order(Group(a,b,c));
24
```

The area marked with dashed lines serves as the fundamental domain, and the
reflections over the edges of the fundamental domain generate $G_{(3,3)}$. The element
$a = (3, 4)$ corresponds to the reflection through the plane $E_a$ containing the edge
from 1 to 2 and passing through the midpoint of the line from 3 to 4. The dual of the
decomposition of the tetrahedron cut along the planes $E_a$, $E_b$ and $E_c$ is the Cayley
graph. From this, we derive, again by gaining an overview of all closed paths in the
Cayley graph, the following presentation.

**Theorem 7.20**   $G_{(3,3)} = \langle s_a, s_b, s_c \mid s_a^2, s_b^2, s_c^2, (s_c s_b)^3, (s_a s_b)^3, (s_a s_c)^2 \rangle$.

We thus obtain the following input for GAP:

```
gap> F := FreeGroup("sa","sb","sc");;
gap> sa:=F.1;; sb:=F.2;; sc:=F.3;;
gap> S4:=F/[sa^2, sb^2, sc^2, (sc*sb)^3, (sa*sb)^3,
            (sa*sc)^2];
<fp group on the generators [ sa, sb, sc ]>
gap> Order(S4);
24
gap> Elements(S4);
[ <identity ...>, sa, sb, sa*sb, sb*sa, sa*sb*sa, sc, sa*sc,
  sb*sc, sa*sb*sc, sb*sa*sc, sa*sb*sa*sc, sc*sb, sa*sc*sb,
  sb*sc*sb, sa*sb*sc*sb, sb*sa*sc*sb, sa*sb*sa*sc*sb,
  sc*sb*sa, sa*sc*sb*sa, sb*sc*sb*sa, sa*sb*sc*sb*sa,
  sb*sa*sc*sb*sa, sa*sb*sa*sc*sb*sa ]
```

We can also ask for the subgroup of orientation-preserving isometries, i.e., the
group $G_{(3,3)}^{+}$. This group is generated by rotations. The rotation around the axis
through the vertex 2 and the center of the opposite triangle by 120° is obtained by
$d_2 = s_a s_b$. The rotation around the axis through the vertex 4 and the center of the
opposite triangle by 120° is obtained by $d_4 = s_b s_c$, as can be seen from Fig. 7.5.
There is also a 180° rotation of the form $d_k = s_c s_a$. From this follow the relations

$d_2^3, d_4^3, d_k^2$. The fundamental domain here is the union of 2 adjacent triangles of Fig. 7.5. It should be clear that $d_2d_4d_k$ is another relation. We show with GAP that this set of relations is sufficient, i.e., that

$$G_{(3,3)}^+ = \langle d_2, d_4, d_k \mid d_2^3, d_4^3, d_k^2, d_2d_4d_k \rangle$$

holds, by noting that the group presented in this way has half as many elements as the group $G_{(3,3)}$ (which was shown in Theorem 3.43):

```
gap> f := FreeGroup("d2","d4","dk");;
gap> d2:=f.1;; d4:=f.2;; dk:=f.3;;
gap> G:=f/[d2^3, d4^3, dk^2, d2*d4*dk];
<fp group on the generators [ d2, d4, dk ]>
gap> Order(G);
12
```

It is the group $A_4$, because the representation of each rotation of the tetrahedron as a permutation of the vertices is always an even permutation. For example, if we rotate the tetrahedron of Fig. 7.5 around the axis through the vertex 1 and the center of the triangle 2,3,4, we obtain the permutations (2,3,4) and (2,4,3) of the vertices, both of which are even.

**The Decompositions of Type** $(n, 2)$  We decompose the sphere $S$ into regular $n$-gons, with always two $n$-gons meeting at a vertex. Such a decomposition is given in Fig. 7.6 in the case $n = 8$.

We conclude, analogous to the case $(3, 3)$:

**Theorem 7.21**  $G_{(n,2)} = \langle s_a, s_b, s_c \mid s_a^2, s_b^2, s_c^2, (s_cs_b)^2, (s_as_b)^n, (s_as_c)^2 \rangle$.

By counting the images of the fundamental region (after drawing the mirror lines) in Fig. 7.6 or by using GAP we see that the order of $G_{(n,2)}$ is equal to $4n$.

*Dual* to this we decompose $S$ into the decomposition of type $(2, m)$, which comprises regular 2-gons, of which always $m$ meet at one vertex. The 2-sphere is thus divided like an orange into slices. The same isometries are possible as in the case $(m, 2)$, and we get exactly the same group presentation.

The decomposition $G_{(4,3)}$ of $S$ corresponds to the cube. Since a cube has 6 sides and $D_4$ with 8 elements is the stabilizer of one side, it follows from the orbit formula that $|G_{(4,3)}| = 6 \cdot 8 = 48$. If you draw the mirror axes of a square on one of the sides

**Fig. 7.6**  Decomposition of $S^2$ of type $(n, 2)$

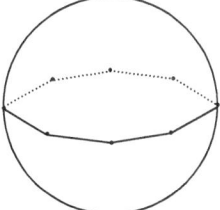

of a cube, you get 8 triangles. Any of these serves as a fundamental region, and we see:

**Theorem 7.22** $G_{(4,3)} = \langle s_a, s_b, s_c \mid s_a^2, s_b^2, s_c^2, (s_c s_b)^4, (s_a s_b)^3, (s_a s_c)^2 \rangle$.

The decomposition of type $(3, 4)$ of $S$ corresponds to the octahedron and has the same group presentation as $G_{(4,3)}$. The *octahedron* consists of 8 regular triangles, of which always 4 meet at one vertex.

The group of orientation-preserving isometries of the cube is just the group $S_4$. To see this, note that every orientation-preserving isometry permutes the 4 diagonals of the cube and two different isometries of the cube permute the diagonals in different ways. Since there are four diagonals, such an isometry is thus representable as an element of $S_4$. Since $S_4$ has 24 elements and the group of orientation-preserving isometries of the cube also has 24 elements, the two groups must be the same.

The Cayley graph of the group $S_4$ is shown in Fig. 4.10 on page 85. Let $P$ be a point on an edge of a cube $W$ near a vertex. If we consider the orbit of $P$ under the group of orientation-preserving isometries of the cube $G_{(4,3)}^+ = S_4$, the image looks very similar to Fig. 4.10. The generators $(1, 2)$, $(2, 4, 3)$ correspond to rotations of the cube, where the cube diagonals have labels $1, 2, 3, 4$. Around each vertex of the cube, there is a small triangle of elements of the orbit of $P$, where the vertices are transformed into each other by conjugates of a rotation of $(2, 4, 3)$. The rotation $(2, 4, 3)$ corresponds to a rotation of $120°$ around the cube diagonal 1. The generator $(1, 2)$ corresponds to a rotation of the cube around an axis that goes through the centers of two opposite cube edges. These cube edges connect the diagonals 1 and 2.

The dodecahedron of Fig. 4.1 on page 65 has as its symmetry group the group $G_{(5,3)}$ with the presentation

$$G_{(5,3)} = \langle s_a, s_b, s_c \mid s_a^2, s_b^2, s_c^2, (s_c s_b)^5, (s_a, s_b)^3, (s_a s_c)^2 \rangle.$$

The orbit formula also easily tells us here $|G_{(5,3)}| = 10 \cdot 12 = 120$. Dual to this, the symmetry group of the *icosahedron* has the same group presentation. It consists of 20 regular triangles.

**Theorem 7.23** *The icosahedron group has the following presentation:*

$$G_{(3,5)} = \langle s_a, s_b, s_c \mid s_a^2, s_b^2, s_c^2, (s_c s_b)^5, (s_a s_b)^3, (s_a s_c)^2 \rangle.$$

It can be proven that the group of orientation-preserving isometries of the dodecahedron is isomorphic to the group $A_5$ (see [AF09]). There are indeed 5 ways to inscribe a cube into a dodecahedron so that each vertex of the cube lies on a vertex of the dodecahedron. These five cubes are permuted by an isometry of the dodecahedron, and a given permutation of the cubes determines in turn an orientation-preserving isometry. Therefore, the group $G_{(3,5)}^+$ is a subgroup of the group $S_5$. Certain rotations realize all cycles of length 3, and these generate the group $A_5$.

Overall, we have calculated here the presentations of the symmetry groups of all *regular polyhedra*. Regular polyhedra consist only of congruent regular $n$-gons, of which always the same number meet at a vertex. They are also called *platonic solids*, because they were already known to PLATO (427–347 BC).

**Theorem 7.24** *If $G$ is the symmetry group of the cube or the icosahedron, then $G = G^+ \times \mathbb{Z}_2$.*

**Proof** This is an application of Theorem 6.3: The point reflection $p$ through the center of the cube or icosahedron forms together with the identity element a subgroup of $G$ isomorphic to $\mathbb{Z}_2$, which has only the identity element in common with $G^+$. In the appendix on matrices, we prove that the point reflection commutes with every other group element. $G^+$ has index 2 in $G$ and is therefore a normal subgroup, as we have proven in Exercise 3. of Sect. 3.4. Every element of $G$ is either from $G^+$, if it is orientation-preserving, or from $pG^+$, if it is orientation-reversing. Thus, the three conditions of Theorem 6.3 are shown, and $G$ can therefore be written as the corresponding direct product. □

## Exercises

1. Prove that if $G$ is the symmetry group of a decomposition $T$ and $G$ acts flag-transitively on $T$, then $T$ is a decomposition into regular $n$-gons, of which always $m$ meet at a vertex.
2. Find presentations for the subgroups of orientation-preserving isometries for the groups $G_{(n,2)}$, $G_{(4,3)}$ and $G_{(3,5)}$.
3. Play with mirrors as in Exercise 7. of Sect. 4.6, and thus create images of regular polyhedra in mirrors. However, you will only see parts of them.
4. Prove the following statement: The group of orientation-preserving isometries of the dodecahedron or icosahedron is isomorphic to the group $A_5$.

## 7.6  Counting Orbits

Anna has a large box of wooden cubes. She wants to paint each side of some cubes red or blue. How many cubes does she need if she wants to consider all possible colorings of the cubes? The cube has six sides and each side can be painted with two colors, which results in $2^6 = 64$ possibilities. However, this does not take into account that cubes can be rotated. If five sides of a cube are blue and only one is red, it doesn't matter which side is red. By rotating the cube, any side can be made the red side.

The group of orientation-preserving isometries of the cube $W^+$ operates on the $2^6$ differently colored cubes. We call this set of cubes $X$. If two cubes are in the same orbit with respect to this operation, then the two cubes can be transformed into each other by a rotation. The number of possible colorings of the wooden cubes is therefore equal to the number of orbits of this operation.

Let $X$ be a $G$-set. Recall that $x \in X$ is called a *fixed point* of $g \in G$ if $g(x) = x$. We define $X^g = \{x \in X \mid g(x) = x\}$.

**Theorem 7.25** *Let $G$ be a finite group that operates on the finite set $X$. The number of different orbits of this operation is*

$$\frac{1}{|G|} \sum_{g \in G} |X^g|.$$

The number of orbits is thus precisely the average number of fixed points.

**Proof** We have

$$\sum_{g \in G} |X^g| = \sum_{x \in X} |G(x)|. \tag{7.2}$$

We are counting in two ways the number of pairs $g, x$ for which $g(x) = x$ holds. On the left-hand side, we count for each group element the elements $x$ that it maps onto itself. On the right-hand side, we count for each element of $X$ the group elements that leave it fixed.

Let $X_1, \ldots, X_n$ be the different orbits. Orbits are disjoint, so

$$\sum_{x \in X} |G(x)| = \sum_{k=1}^{n} \sum_{x \in X_k} |G(x)|. \tag{7.3}$$

According to Theorem 4.30, points from the same orbit have conjugate stabilizers, and these stabilizers are all of the same size. That is, if $y \in X_k$, then

$$\sum_{x \in X_k} |G(x)| = |X_k| \cdot |G(y)| = |Gy| \cdot |G(y)|.$$

According to the orbit-stabilizer theorem (Theorem 4.35), $|Gy| \cdot |G(y)| = |G|$. By Eq. (7.3), we get

$$\sum_{x \in X} |G(x)| = \sum_{k=1}^{n} |G| = n|G|.$$

From (7.2), we obtain

$$\sum_{g \in G} |X^g| = n|G|,$$

and this corresponds to the assertion. □

**Example 7.26** The group $D_4$ operates on the vertices $X$ of a square. Because the operation is transitive, there is only one orbit, and Theorem 7.25 tells us

$$8 = |D_4| = \sum_{g \in D_4} |X^g|.$$

The two reflections on the diagonals of a square each have two fixed points, and the identity has 4, making a total of the required 8. No other element of $D_4$ has fixed points.

**Example 7.27** Let $G$ be the stabilizer of a face $S$ in the symmetry group of the cube. It holds that $G \cong D_4$. $G$ operates on the set of edges $K$ of the cube with 3 orbits. The boundary edges of $S$ form one orbit. If $S'$ is the face opposite $S$, then the boundary edges of $S'$ form a second orbit. The remaining 4 edges form the third orbit.

Theorem 7.25 gives us

$$3 = \frac{1}{|G|} \sum_{g \in G} |K^g| = \frac{1}{8} \sum_{g \in G} |K^g|.$$

So $\sum |K^g| = 24$. The identity leaves 12 edges fixed. A reflection of $G$ through a plane passing through opposite edge centers of $S$ leaves 4 edges fixed, and a reflection through a plane passing through opposite vertices of $S$ leaves 2 edges fixed. In total: $24 = 12 + 2 \cdot 4 + 2 \cdot 2$.

Back to our initial example. Conjugate group elements have the same number of fixed points (see Theorem 4.27). So Anna has to take one element of each conjugacy class of elements of the group $W^+$ and check how many colored cubes this element leaves fixed. With the vertex labels of Fig. 1.8, $a = (1, 5, 6, 2)(4, 8, 7, 3)$ is a 90° rotation about an axis through opposite face centers, $b = (2, 7, 5)(1, 3, 8)$ is a 120° rotation about a cube diagonal and $c = (1, 5)(3, 7)(2, 8)(6, 4)$ is a 180° rotation about a line through the centers of edges 1, 5 and 3, 7. These three elements generate $W^+$:

```
gap> Wn:=Group((1,5,6,2)(4,8,7,3),(2,7,5)(1,3,8),
    (1,5)(3,7)(2,8)(6,4));
Group([ (1,5,6,2)(3,4,8,7), (1,3,8)(2,7,5),
        (1,5)(2,8)(3,7)(4,6) ])
gap> Order(Wn);
24
gap> C:=ConjugacyClasses(Wn);
[ ()^G, (2,4,5)(3,8,6)^G, (1,2)(3,5)(4,6)(7,8)^G,
  (1,2,3,4)(5,6,7,8)^G, (1,3)(2,4)(5,7)(6,8)^G ]
gap> List(C, i->Order(i));
[ 1, 8, 6, 6, 3 ]
```

The conjugacy classes are therefore, in the order output by GAP:

1. the identity as its own class,
2. 8 rotations by 120 or 240° about cube diagonals,
3. 6 rotations by 180° about lines through opposite edge centers,
4. 6 rotations by 90 or 270° about lines through opposite face centers,
5. 3 rotations by 180° about lines through opposite face centers.

The identity leaves all $2^6$ cubes fixed. The 120- and 240° rotations each leave $2^2 = 4$ cubes fixed. If the diagonal about which such a rotation occurs, for example, goes through vertices 1 and 7, then one must color all cube faces that have the vertex 1 in their boundary with the same color and all that have the vertex 7 in their boundary with the same color. This gives 4 colorings. The 180° rotations about opposite edge centers and also the 90- or 270° rotations each leave $2^3$ cubes fixed. The rotations about opposite face centers each leave $2^4 = 16$ cubes fixed. In total, we therefore have:

$$\frac{1}{24}\left((1 \cdot 2^6) + (8 \cdot 2^2) + (6 \cdot 2^3) + (6 \cdot 2^3) + (3 \cdot 2^4)\right) = 10.$$

So, Anna needs 10 wooden cubes to be able to create all possible colorings.

For this simple problem, one could have counted the number of cubes in a different way, but for problems with more possibilities, this is no longer an option, and one must use Theorem 7.25. One could ask, for example, how many different ways one can color the faces of an icosahedron with three colors. An icosahedron has 20 triangular faces, so there are $3^{20}$ colorings, without considering rotations. That's over 3 billion.

## Exercises

1. Color each edge of a cube blue or red. How many different such cube colorings are there?
2. A round cake is cut into 8 equal pieces. In how many ways can you place red and green candles so that exactly one candle is in the middle of each piece?
3. Assume that a finite group $G$ acts transitively on a finite set $X$, where $|X| > 1$. Show with the help of Theorem 7.25 that there must be an element $g \in G$ that has no fixed point in $X$.

# Chapter 8
# Abelian and Solvable Groups

We begin this chapter by looking at commutators and the commutator subgroup of a group. The commutator subgroup of a non-abelian group is a normal subgroup that is not the trivial group. In the second section, on abelian groups, a classification (i.e. a complete list) of finitely generated abelian groups is given. The last section introduces solvable groups, which are important in connection with field extensions in Galois theory. We do not go into field extensions here, but solvable groups are also interesting within group theory.

## 8.1 Commutators

If $G$ is a group and $g, h \in G$, then $[g, h] = ghg^{-1}h^{-1}$ is called the *commutator* of $g$ and $h$. $g$ and $h$ *commute* in $G$ if $[g, h] = 1$. In an abelian group, any two elements commute, and only the identity element is a commutator. If a group is not abelian, there are commutators that are not the identity element.

If $J$ and $H$ are subgroups of a group $G$, we write

$$[J, H] = \langle [j, h] \mid j \in J, h \in H \rangle$$

for the subgroup of $G$ generated by all commutators $[j, h]$ ($j \in J, h \in H$).

Let $G$ be a group. The *commutator subgroup* $G' = [G, G]$ is the subgroup generated by all commutators:

$$[G, G] = \{ g_1 h_1 g_1^{-1} h_1^{-1} g_2 h_2 g_2^{-1} h_2^{-1} \ldots g_n h_n g_n^{-1} h_n^{-1} \mid g_i, h_i \in G \}.$$

Its elements thus consist of products of commutators of $G$. Not every product of commutators is a commutator, although explicit examples are hard to find (see Fischer [Fis17]).

S. Rosebrock, *Visual Group Theory*, Springer Undergraduate Mathematics Series, https://doi.org/10.1007/978-3-662-69365-0_8

The commutator subgroup of an abelian group is the trivial group. If $G$ is not abelian, the commutator subgroup is nontrivial.

If $G$ is the symmetry group of a figure, then only orientation-preserving isometries are contained in the commutator subgroup $G'$ because each commutator contains an even number of orientation-reversing isometries. For each orientation-reversing isometry $g$, $g^{-1}$ also appears in the commutator.

We calculate the commutator subgroup of the dihedral group $D_n$ of the regular $n$-gon. If $s$ is a reflection and $d$ is the rotation by $360/n$ degrees, then the relation $sd^k = d^{n-k}s$ holds in $D_n$ (see formula (6.4) on page 123). $[d^k, s]$ is an element of the commutator subgroup $D'_n$. We have

$$[d^k, s] = d^k s d^{-k} s^{-1} = d^k s d^{n-k} s = d^k s s d^k = d^k d^k = d^{2k},$$

because $d^{n-k} = d^{-k}$ and $s = s^{-1}$. All even powers of $d$ are thus commutators. There are no more nontrivial commutators. The commutator subgroup thus consists of

$$D'_n = \{id, d^2, d^4, \dots, d^{2(n-1)}\}.$$

If $n$ is even, then $D'_n$ consists of every second rotation, for example $D'_6 = \{id, d^2, d^4\}$, and $D'_n$ is isomorphic to $\mathbb{Z}_{n/2}$. For odd $n$, $D'_n$ consists of all rotations $D'_n = D^+_n$ and is isomorphic to $\mathbb{Z}_n$.

We verify this with GAP. `DerivedSubgroup` gives us the commutator subgroup. For the group $D_5$ we get the cyclic group of order 5 as the commutator subgroup.

```
gap> D5:=DihedralGroup(10);
<pc group of size 10 with 2 generators>
gap> G5:=DerivedSubgroup(D5);
Group([ f2 ])
gap> Order(G5);
5
```

The commutator subgroup of the group $D_6$ yields the cyclic group of order 3.

```
gap> D6:=DihedralGroup(12);
<pc group of size 12 with 3 generators>
gap> G6:=DerivedSubgroup(D6);
Group([ f3 ])
gap> Order(G6);
3
```

The commutator subgroup $G' = [G, G]$ is even a normal subgroup of the group $G$. Indeed, if $h \in G'$ and $g \in G$, then $ghg^{-1} \in G'$. According to Lemma 3.38, this is sufficient to prove $G'$ is a normal subgroup. If $h \in G'$, then $ghg^{-1} = (ghg^{-1}h^{-1}) \cdot h$, and because $ghg^{-1}h^{-1} \in G'$ and $h \in G'$, it follows that $(ghg^{-1}h^{-1}) \cdot h \in G'$.

Since $G'$ is a normal subgroup, we can form the factor group $G/G'$. $G/G'$ is an abelian group, because every commutator is quotiented to the identity element. Every commutator in $G/G'$ is the identity element. $G/G'$ is called the *abelianization* of $G$. The commutator subgroup is the smallest normal subgroup such that the quotient becomes abelian.

**Theorem 8.1**  *The abelianization of the free group $F_n$ of rank $n$ is $\mathbb{Z}^n$.*

**Proof**  Let $F_n = \langle x_1, \ldots, x_n \rangle$. The factor group $F_n/[F_n, F_n]$ is abelian. Every element of $F_n/[F_n, F_n]$ can therefore be represented in exactly one way in the form

$$x_1^{m_1} \ldots x_n^{m_n}[F_n, F_n].$$

The mapping

$$x_1^{m_1} \ldots x_n^{m_n}[F_n, F_n] \to (m_1, \ldots, m_n)$$

is therefore an isomorphism between $F_n/[F_n, F_n]$ and $\mathbb{Z}^n$.                     □

A group is called *perfect* if it is equal to its commutator subgroup. In other words, a group is perfect if every one of its elements can be written as a commutator. A non-abelian simple group $G$ is perfect, because if $G$ is simple, then it has no proper normal subgroups. Since the commutator subgroup is nontrivial, because $G$ is non-abelian, and also normal, it must be equal to the whole group. Therefore, according to Theorem 4.41, the alternating group $A_5$ is perfect.

We calculate the commutator subgroup of the symmetric group $S_n$ for $n \geq 3$. We have

$$[(1, 2), (1, 3)] = (1, 2) \circ (1, 3) \circ (1, 2)^{-1} \circ (1, 3)^{-1} = (1, 2) \circ (1, 3) \circ (1, 2) \circ (1, 3)$$

$$= ((1, 2) \circ (1, 3))^2 = (1, 3, 2)^2 = (1, 2, 3).$$

So $(1, 2, 3)$ is a commutator, and because the numbers in the calculation are arbitrarily interchangeable, every 3-cycle in $S_n$ is a commutator. Because the alternating group $A_n$ is generated by 3-cycles (see Theorem 4.8), $A_n$ is therefore a subgroup of the commutator subgroup $S_n'$.

In fact, $A_n = S_n'$, because a transposition $(i, j)$ is never a commutator. $S_n$ is the symmetry group of an $n - 1$-dimensional tetrahedron in $\mathbb{R}^{n-1}$ (see Exercise 2. of Sect. 4.1), and since $(i, j)$ describes a reflection, it is therefore orientation-reversing. So $S_n' \neq S_n$. The order of $S_n'$ must be a divisor of the group order $|S_n| = n!$, and at least $|A_n| = n!/2$. Since $S_n'$ contains the group $A_n$, it must be equal to $A_n$. Of course, one also sees directly that every commutator must be an even permutation, because it is composed of an even number (namely 4) of permutations. So we have proven:

**Theorem 8.2**  *For all $n \geq 3$, $S_n' = A_n$.*

## Exercises

1. Let $G = \langle X \mid R \rangle$. Show that $G/G'$, the abelianized group, has the presentation $\langle X \mid R, [X, X] \rangle$. To the presentation of $G$, all commutators of generators are therefore added.
2. Show that the commutator subgroup $Q'$ of the quaternion group $Q$ (see Exercise 6. of Sect. 5.1) is $\{\pm 1\}$. Show that $Q/Q' \cong \mathbb{Z}_2 \times \mathbb{Z}_2$.
3. Prove that $[x, y] = [y, x]^{-1}$.
4. Determine the commutator subgroup of $\mathcal{E}$, the symmetry group of the Euclidean plane.

## 8.2   Abelian Groups

A group of the form

$$\mathbb{Z}_{n_1} \times \mathbb{Z}_{n_2} \times \cdots \times \mathbb{Z}_{n_j} \times \mathbb{Z}^k$$

is abelian, because each factor is abelian. It is also finitely generated, because the group can be generated by $j$ generators for the finite factors and $k$ generators for $\mathbb{Z}^k$.

Finitely generated abelian groups can be classified, and we want to prove this classification here. In the following classification theorem, the symbol | means *is a divisor of*. For example, $3|15$.

**Theorem 8.3** *Every finitely generated abelian group is isomorphic to a direct product of the form*

$$\mathbb{Z}_{n_1} \times \mathbb{Z}_{n_2} \times \cdots \times \mathbb{Z}_{n_j} \times \mathbb{Z}^k. \tag{8.1}$$

*Here, $k$ or $j$ may be zero, and the $n_i$ are natural numbers $n_i \geq 2$ such that $n_i | n_{i+1}$.*

A finite abelian group is thus isomorphic to a direct product of finite cyclic groups, and a finitely generated abelian group without elements of finite order is thus isomorphic to $\mathbb{Z}^k$, a direct product of $k$ $\mathbb{Z}$-factors.

We have already seen in Exercise 4. of Sect. 6.1 that $\mathbb{Z}_n$ is isomorphic to $\mathbb{Z}_p \times \mathbb{Z}_q$ if $n = p \cdot q$ and $p, q$ are coprime. This is consistent with Theorem 8.3. But what if $p, q$ are not coprime, but $p$ does not divide $q$? Then this also fits into our theorem. For example, we have

$$\mathbb{Z}_6 \times \mathbb{Z}_9 \cong \mathbb{Z}_2 \times \mathbb{Z}_3 \times \mathbb{Z}_9 \cong \mathbb{Z}_3 \times \mathbb{Z}_{18}.$$

We can also apply this to subgroups of non-finitely generated abelian groups. Consider the subgroup $U = \langle 2, \sqrt{2} \rangle$ of $(\mathbb{R}, +)$. Because a multiple of $\sqrt{2}$ is never an integer, according to Theorem 8.3 we have $U \cong \mathbb{Z}^2$.

A generating system $S$ of a group $G$ is called *minimal* if $G$ does not contain a generating system with fewer elements than $S$.

***Proof of Theorem 8.3*** Let $G$ be any finitely generated abelian group. Let $\{g_1, \ldots, g_t\}$ be a minimal generating system of $G$. Each element $w \in G$ can be written as

$$w = g_1^{m_1} \ldots g_t^{m_t}$$

because $G$ is abelian. If

$$1 = g_1^{m_1} \ldots g_t^{m_t}$$

is a relation in $G$ only if $m_1 = m_2 = \cdots = m_t = 0$, then $G \cong \mathbb{Z}^t$. There is then an isomorphism $\phi \colon G \to \mathbb{Z}^t$, by setting

$$\phi(g_1^{m_1} \ldots g_t^{m_t}) = (m_1, m_2, \ldots, m_t).$$

From now on, we assume that there are further relations in $G$. Among the relations of all minimal generating systems, let $r_1$ be the smallest occurring exponent. We assume that

$$1 = g_1^{r_1} g_2^{m_2} \ldots g_t^{m_t} \tag{8.2}$$

is such a relation and that $r_1$ is the exponent of the generator $g_1$. If $r_1$ appears as the exponent of another generator, rename the generators. $r_1$ must be a divisor of $m_2$. To see this, we divide $m_2$ by $r_1$ with remainder, $m_2 = qr_1 + z$ where $0 \le z < r_1$, and show that $z = 0$ must hold. Equation (8.2) becomes

$$1 = g_1^{r_1} g_2^{qr_1+z} g_3^{m_3} \ldots g_t^{m_t} = (g_1 g_2^q)^{r_1} g_2^z g_3^{m_3} \ldots g_t^{m_t}.$$

The set $\{(g_1 g_2^q), g_2, g_3, \ldots, g_t\}$ is also a minimal generating system, because every group element $g_1^{m_1} \ldots g_t^{m_t} \in G$ can be written in the new generators: $(g_1 g_2^q)^{m_1} g_2^{m_2-qm_1} g_3^{m_3} \ldots g_t^{m_t}$. If $z \neq 0$, then, because $z < r_1$, we have a contradiction to the choice of $r_1$ as the smallest occurring exponent. $z$ appears in the new generating system as an exponent of $g_2$. Therefore, $z = 0$ and hence $m_2 = qr_1$. Similarly, we show that $r_1$ divides each of the numbers $m_3$ to $m_t$, and we set $q_i = m_i/r_1$ for $3 \le i \le t$.

We now change the generating system. We set $h_1 = g_1 g_2^q g_3^{q_3} \ldots g_t^{q_t}$. Our new generating system is $h_1, g_2, \ldots, g_t$. The relation (8.2) now reads

$$1 = h_1^{r_1}. \tag{8.3}$$

This can be seen by replacing the element $h_1$ in Eq. (8.3) by

$$g_1 g_2^q g_3^{q_3} \cdots g_t^{q_t}.$$

$r_1$ was initially chosen as the smallest exponent occurring in a relation, so no smaller positive exponent of $h_1$ is the identity, and thus $h_1$ has order $r_1$.

Let $U = \langle h_1 \rangle < G$ and $G_1 = \langle g_2, \ldots, g_t \rangle < G$. $U$ is isomorphic to $\mathbb{Z}_{r_1}$. It holds that $U G_1 = G$ and $U \cap G_1 = \{e\}$. Because $G$ is abelian, $U$ and $G_1$ are normal subgroups of $G$. By Theorem 6.3, it follows that

$$G = U \times G_1 \cong \mathbb{Z}_{r_1} \times G_1.$$

Now we do exactly the same with $G_1$ as we did before with $G$. So either $G_1 = \mathbb{Z}^{t-1}$, and we are done, or $G_1 = \mathbb{Z}_{r_2} \times G_2$ and $G \cong \mathbb{Z}_{r_1} \times \mathbb{Z}_{r_2} \times G_2$. In the second case, $r_2$ must appear as an exponent in a relation, so

$$1 = x_2^{r_2} x_3^{m_3'} \cdots x_t^{m_t'}.$$

Together with the relation (8.3), we have a relation

$$1 = h_1^{r_1} x_2^{r_2} x_3^{m_3'} \cdots x_t^{m_t'}$$

in $G$. Above, however, we saw that in any such relation $r_1$ is a divisor of $r_2$. We continue our proof with $G_2$ as before with $G_1$, then with $G_3$ etc. After at most $t$ steps, the process ends, and the theorem is proven.                                           $\square$

It can also be shown that any two such groups are different:

**Theorem 8.4** Let $G_1 = \mathbb{Z}_{n_1} \times \cdots \times \mathbb{Z}_{n_t} \times \mathbb{Z}^k$ with $n_i | n_{i+1}$ and $G_2 = \mathbb{Z}_{m_1} \times \cdots \times \mathbb{Z}_{m_s} \times \mathbb{Z}^p$ with $m_i | m_{i+1}$. If $G_1$ is isomorphic to $G_2$, then it follows that $t = s$, $k = p$ and $n_i = m_i$ for $1 \le i \le t$.

A proof can be found, for example, in Armstrong [Arm88]. This shows, for instance, that $\mathbb{Z}_9$ is not isomorphic to $\mathbb{Z}_3 \times \mathbb{Z}_3$.

Direct products of abelian groups are not unique. An example is the following:

**Example 8.5** Let $U = \langle x \rangle \cong \mathbb{Z}_2$, where the order of $x$ is 2 and $V = \langle y \rangle \cong \mathbb{Z}$. Let

$$G = U \times V = \{(x^n, y^m) \mid n, m \in \mathbb{Z}\} \cong \mathbb{Z}_2 \times \mathbb{Z}.$$

Instead of the subgroup $V = \{(e, y^m) \mid m \in \mathbb{Z}\}$ ($e$ being the identity element in $U$), we consider the subgroup $H < G$ defined by $H = \{(x^m, y^m) \mid m \in \mathbb{Z}\}$.

It holds that $G = U \times H$ according to Theorem 6.3: In fact, $G = UH$, because for $(x^n, y^m) \in G$ with $n \equiv m \mod 2$, $(x^n, y^m) \in V$, and otherwise $(x^n, y^m) = (x, e)(x^{n-1}, y^m)$, where $(x^{n-1}, y^m) \in V$. $(x, e) \notin V$, and therefore $U \cap V =$

$(e, e)$ is the identity element of $G$. The group $G$ can therefore be represented in two different ways as a direct product.

## Exercises

1. Let $p_1, p_2, \ldots, p_n$ be different prime numbers. Show that an abelian group of order $p_1 \cdots p_n$ must be cyclic.
2. Let $U = \langle (1, 1) \rangle$ be a subgroup of $\mathbb{Z}_p \times \mathbb{Z}_q$. What is the order of $U$? To which group is $U$ isomorphic?

## 8.3 Solvable Groups

Let $G$ be a group. The *higher commutator groups* are defined as follows:

$$G^{(0)} = G, \quad G^{(1)} = [G, G], \quad G^{(n+1)} = [G^{(n)}, G^{(n)}].$$

For example, $G^{(2)}$ is the subgroup of $G$ formed by all commutators of commutators. $G^{(i)}$ is a normal subgroup of $G^{(i-1)}$, because $G^{(i)}$ is the commutator subgroup of $G^{(i-1)}$ (see Sect. 8.1).

We thus obtain a descending chain of subgroups, the so-called *commutator series*:

$$G = G^{(0)} \rhd G^{(1)} \rhd G^{(2)} \rhd G^{(3)} \rhd \cdots$$

For every group $H$, $H/[H, H]$ is abelian, i.e., the factor groups $G^{(i)}/G^{(i+1)}$ are all abelian groups.

**Definition 8.6** A *normal series* of a group $G$ is a descending chain of normal subgroups

$$G = N_0 \rhd N_1 \rhd N_2 \rhd \cdots \rhd N_k = \{e\}.$$

The factor groups $N_i/N_{i+1}$ are called *factors* of the normal series.

The important point here is that we end with the trivial group, which is not necessarily the case with descending commutator subgroups.

There is an analogy to natural numbers: For each $n \in \mathbb{N}$ there is a chain of natural numbers

$$n = n_0 > n_1 > \cdots > n_{k-1} > n_k = 1 \tag{8.4}$$

such that the quotients $n_i/n_{i+1}$ are prime. Prime numbers are the building blocks of natural numbers. Finite groups are made out of simple groups in an analogous way:

**Theorem 8.7** *Each finite group G has a normal series*

$$G = N_0 \triangleright N_1 \triangleright N_2 \triangleright \cdots \triangleright N_k = \{e\}$$

*with simple factors.*

***Proof*** Let $G$ be a non-trivial group and start with $G = N_0 \triangleright \{e\}$. Assume inductively we have constructed $G = H_0 \triangleright H_1 \triangleright H_2 \triangleright \cdots \triangleright H_m = \{e\}$ and $H_{i-1}/H_i$ is not simple. Then take a normal subgroup $K/H_i$ in $H_{i-1}/H_i$ which is different from the whole group and from $H_i$ and place $K$ in the already generated normal series between $H_{i-1}$ and $H_i$: $H_{i-1} \triangleright K \triangleright H_i$. Repeat this process until all factors are simple.                                                                                         $\square$

Regarding the natural numbers, one can reconstruct $n$ from the factors $n_i/n_{i+1}$ in (8.4). This is not possible in normal series of groups, as can be seen in this simple example:

$$\mathbb{Z}_4 \triangleright \mathbb{Z}_2 \triangleright \{0\} \qquad D_2 \triangleright \{id, d\} \triangleright \{id\},$$

where $D_2$ is the Klein four-group, here as the symmetry group of the rhombus, with a rotation $d$ of $180°$ about the midpoint of the rhombus. In both normal series the factors are isomorphic to $\mathbb{Z}_2$.

As examples of normal series, we have:

$$S_n \triangleright A_n \triangleright \{e\}, \quad \forall n \geq 5.$$

In Theorem 8.2 we saw that $S'_n = A_n$ for $n \geq 3$. In addition, we have:

$$S_4 \triangleright A_4 \triangleright V \triangleright \{e\},$$

where $V$ is the Klein four-group. If $A_4$ are the even permutations of the set $\{1, 2, 3, 4\}$, then we have

$$V = \{id, (1, 2)(3, 4), (1, 3)(2, 4), (1, 4)(2, 3)\}.$$

$V$ consists of even permutations, so that $V < A_4$ holds. But $V$ is even normal in $A_4$, because

$$(1, 2, 3) \circ (1, 2)(3, 4) \circ (1, 3, 2) = (1, 4)(2, 3).$$

Conjugates of elements of $V$ are therefore again in $V$. According to Lemma 3.38, this is already sufficient to show that $V$ is normal in $A_4$.

**Definition 8.8** A group $G$ is called *solvable* if $G$ has a normal series with abelian factors.

The name "solvable group" comes from the relationship to the solvability of algebraic polynomial equations. We will not go into this here.

Every abelian group $G$ is solvable because $G \rhd \{e\}$.

**Lemma 8.9** *Let $G$ be a group. There is a $k \in \mathbb{N}$ with $G^{(k)} = \{e\}$ if and only if $G$ is solvable.*

**Proof** If there is a $k \in \mathbb{N}$ with $G^{(k)} = \{e\}$, then $G$ is solvable. Take the commutator series.

For the converse, let

$$G = N_0 \rhd N_1 \rhd N_2 \rhd \cdots \rhd N_k = \{e\}$$

be a normal series with abelian factors. We show

$$G^{(i)} < N_i \text{ for all } 0 \le i \le k. \tag{8.5}$$

Then, because $N_k = \{e\}$ and $G^{(k)} < N_k$, the condition $G^{(k)} = \{e\}$ is fulfilled.

For $i = 0$, we have $G < G$, which is true. By induction, we assume that (8.5) is proven for $i$. $N_i/N_{i+1}$ is abelian by assumption, and because the commutator group is the smallest normal subgroup that makes a group abelian, we have $[N_i, N_i] < N_{i+1}$. It now follows that

$$G^{(i+1)} = [G^{(i)}, G^{(i)}] < [N_i, N_i] < N_{i+1}$$

where $[G^{(i)}, G^{(i)}] < [N_i, N_i]$ follows from the induction assumption $G^{(i)} < N_i$.

□

**Theorem 8.10** *$S_n$ is solvable if and only if $n \le 4$.*

**Proof** $S_2$ is abelian and therefore solvable. The normal series

$$S_3 \rhd A_3 \rhd \{e\} \text{ and } S_4 \rhd A_4 \rhd V \rhd \{e\}$$

have abelian factors (the orders of the factor groups are prime numbers, and there are only abelian groups of prime order, except $V \rhd \{e\}$), and therefore $S_3$ and $S_4$ are solvable.

For $n \ge 5$, observe that every 3-cycle of $A_n$ is a commutator:

$$(i, j, k) = (i, j, l) \circ (i, k, m) \circ (l, j, i) \circ (m, k, i).$$

Because the group $A_n$ is generated by 3-cycles (see Theorem 4.8), every permutation of $A_n$ is a product of commutators, and therefore $A_n = [A_n, A_n]$. But then $A_n^{(i)} = A_n$ for all $i$, and therefore the commutator series

$$S_n \rhd A_n \rhd A_n \rhd A_n \rhd \cdots$$

does not terminate. Then Lemma 8.9 shows the claim.                                    □

We check with GAP whether the quaternion group is solvable.

```
gap> Q:=SmallGroup(8,4);;
gap> IsSolvable(Q);
true
```

One can prove that the group $A_5$ is the smallest non-solvable group. We consider the group $G = \langle 1 \rangle \cong \mathbb{Z}_{30}$ and two normal series of $G$:

$$G \rhd \langle 10 \rangle \rhd \{e\} \quad \text{and} \quad G \rhd \langle 2 \rangle \rhd \langle 6 \rangle \rhd \{e\}.$$

The first normal series can be refined, i.e., another normal subgroup can be inserted:

$$G \rhd \langle 5 \rangle \rhd \langle 10 \rangle \rhd \{e\}.$$

After this refinement, we consider the factor groups of the two normal series. The first normal series has the following factor groups:

$$G/\langle 5 \rangle \cong \mathbb{Z}_5, \quad \langle 5 \rangle / \langle 10 \rangle \cong \mathbb{Z}_2, \quad \langle 10 \rangle / \{e\} \cong \mathbb{Z}_3.$$

The second normal series has the following factor groups:

$$G/\langle 2 \rangle \cong \mathbb{Z}_2, \quad \langle 2 \rangle / \langle 6 \rangle \cong \mathbb{Z}_3, \quad \langle 6 \rangle / \{e\} \cong \mathbb{Z}_5.$$

So the same factor groups appear, only they are permuted. This motivates the following definition:

**Definition 8.11** Let $G$ be a group. A normal series

$$G = N_0 \rhd N_1 \rhd N_2 \rhd \cdots \rhd N_k = \{e\}$$

is called a *refinement* of a normal series

$$G = H_0 \rhd H_1 \rhd H_2 \rhd \cdots \rhd H_n = \{e\}$$

if each group $H_i$ is contained in $\{N_0, N_1, \ldots, N_k\}$. Two normal series of a group $G$ are called *equivalent* if there is a bijective mapping between their factor groups.

The following refinement theorem of Schreier (1928) is important. A proof can be found in Rotman [Rot95]:

**Theorem 8.12** *Any two normal series of a group have equivalent refinements.*

We prove:

**Theorem 8.13** *Let G be a finite solvable group. Then there is a normal series with cyclic factors of prime order.*

**Proof** We show:

> Let $N \lhd G$, and $N$ be a proper normal subgroup of $G$ (i.e. the factor group is nontrivial). Let $G/N$ be abelian and finite. Then there is a normal subgroup $H \lhd G$ with $N \lhd H$ and $H/N \cong \mathbb{Z}_p$, where $p$ is a prime number.

We then take any normal series of $G$ and refine it until all factors have prime order.

To prove the claim: Because $G/N$ is nontrivial, there is a nontrivial element $g \in G/N$. The group $\langle g \rangle < G/N$ is a cyclic group, and because $G$ is finite, it is isomorphic to a group $\mathbb{Z}_n$.

According to Exercise 8. of Sect. 3.3, for every prime divisor $p|n$ there is a subgroup $U < \langle g \rangle$ such that $U \cong \mathbb{Z}_p$. $G/N$ is abelian, so $U \lhd G/N$.

We consider the surjective homomorphism $\phi\colon G \to G/N$, and we define $H$ as $\phi^{-1}(U)$. According to Lemma 8.14, which we will prove shortly, $H$ is normal in $G$. The following diagram shows the connection:

$$\phi\colon G \to G/N$$
$$\nabla \qquad \nabla$$
$$\phi\colon H \to U \cong \mathbb{Z}_p$$

$H/N \cong U$, because $\phi(H) = U$ and $\phi$ is just the canonical homomorphism. $\qquad \square$

**Lemma 8.14** *Let $\phi\colon G \to J$ be a group homomorphism and $U \lhd J$. Then $\phi^{-1}(U) \lhd G$.*

**Proof** Pre-images of subgroups are subgroups (see Theorem 3.32), i.e. $\phi^{-1}(U) < G$. Let $g \in \phi^{-1}(U)$, so there is a $u \in U$ with $\phi(g) = u$. For any $h \in G$ we have

$$\phi(hgh^{-1}) = \phi(h) \cdot u \cdot \phi(h)^{-1} \in U,$$

because $U$ is normal in $J$. It follows that $hgh^{-1} \in \phi^{-1}(U)$. $\qquad \square$

**Lemma 8.15** *Let $\phi\colon G \to H$ be a group homomorphism. Then $\forall n \in \mathbb{N}$*

$$\phi(G^{(n)}) = \phi(G)^{(n)}.$$

**Proof** We prove the equation by induction over $n$. For $n = 0$, the equation becomes $\phi(G) = \phi(G)$. Let $n > 1$. It holds that $\phi([x, y]) = [\phi(x), \phi(y)]$. From this we get

$$\phi(G^{(n+1)}) = \phi([G^{(n)}, G^{(n)}]) = [\phi(G^{(n)}), \phi(G^{(n)})]$$
$$= [\phi(G)^{(n)}, \phi(G)^{(n)}] = \phi(G)^{(n+1)},$$

because, by the induction assumption,

$$[\phi(G^{(n)}), \phi(G^{(n)})] = [\phi(G)^{(n)}, \phi(G)^{(n)}].$$

$\square$

**Theorem 8.16** *Let $N \lhd G$. $G$ is solvable if and only if $N$ and $G/N$ are solvable.*

**Proof** If $G$ is solvable, then $N$ is also solvable according to Exercise 1. Let $\phi \colon G \to G/N$ be the canonical homomorphism. By Lemma 8.15 we have

$$(G/N)^{(n)} = \phi(G)^{(n)} = \phi(G^{(n)}).$$

So if $G^{(n)} = \{e\}$, then $(G/N)^{(n)} = \{e\}$.

Conversely, let $N$ and $G/N$ be solvable. Let $m \in \mathbb{N}$ such that $N^{(m)}$ and $(G/N)^{(m)}$ are each the trivial group. By Lemma 8.15 $\phi(G^{(m)}) = (G/N)^{(m)} = \{e\}$, and therefore $G^{(m)} < N$. From this follows

$$G^{(2m)} = (G^{(m)})^{(m)} < N^{(m)} = \{e\}.$$

So the commutator series terminates at the trivial group, and therefore $G$ is solvable.

$\square$

**Theorem 8.17** *Every p-group is solvable.*

**Proof** Let $G$ be a group and $|G| = p^n$, where $p$ is a prime number. We prove the theorem by induction over $n$. For $n = 0$ there is nothing to show. We consider the center $C(G) \lhd G$. Because $C(G)$ is a subgroup of $G$, it holds that $|C(G)| = p^m$ with $m \leq n$. According to Theorem 4.44 the center of a $p$-group is nontrivial, i.e. $m \geq 1$. Thus, it follows that $|G/C(G)| = p^{n-m} < p^n$. By the induction assumption, $G/C(G)$ is solvable, $C(G)$ is solvable because it is abelian, and the claim follows from Theorem 8.16.

$\square$

Feit and Thompson showed in 1963 that every finite group of odd order is solvable. The proof is over 200 pages long.

**Definition 8.18** Let $G$ be a group. A *central series* of $G$ is a normal series

$$G = N_0 \rhd N_1 \rhd N_2 \rhd \cdots \rhd N_k = \{e\},$$

where the factors $N_{i-1}/N_i$ lie in the center $C(G/N_i)$. $G$ is called *nilpotent* if it has a central series. The length of the shortest central series of $G$ is its *nilpotency class*.

The condition that the factors $N_{i-1}/N_i$ lie in the center $C(G/N_i)$ is obviously equivalent to the condition $[G, N_{i-1}] < N_i$.

Only the trivial group has nilpotency class 0. If the factors $N_{i-1}/N_i$ lie in the center $C(G/N_i)$, then the factors are naturally abelian, so nilpotent groups are solvable. The converse is generally false. The group $S_3$ is solvable (see Theorem 8.10), but not nilpotent, because its center is trivial (see Theorem 4.45).

Abelian groups are nilpotent with nilpotency class 1, because if $G$ is abelian, then $C(G/N_i) = G/N_i$. If $G$ is a non-abelian group such that $G/C(G)$ is abelian, then $G$ is nilpotent of nilpotency class 2 with central series

$$G \rhd C(G) \rhd \{e\}.$$

**Theorem 8.19** *A p-group is nilpotent.*

**Proof** Let $G$ be a $p$-group with $|G| > 1$. According to Theorem 4.44, the center $C(G)$ of $G$ is nontrivial. Inductively, as in the proof of Theorem 8.17, we can assume that $G/C(G)$ is nilpotent. We consider a central series

$$G/C(G) = N_0 \rhd N_1 \rhd N_2 \rhd \cdots \rhd N_k = \{e\}$$

for $G/C(G)$. There is the canonical homomorphism $\phi: G \to G/C(G)$. The pre-images $\phi^{-1}(N_i)$ form a central series for $G$, if we append the trivial group to the resulting central series at the pre-image $C(G)$ of $N_k$.                                  □

It can be proven that a finite group is nilpotent if and only if it is a direct product of $p$-groups (for different prime numbers $p$).

**Example 8.20** The *Heisenberg group H* is presented by

$$\langle x, y, z \mid [x, z], [y, z], [x, y] = z \rangle.$$

It holds that $[H, H] = \langle z \rangle \cong \mathbb{Z}$. $H/[H, H]$ is abelian, thus it lies in the center $C(H/[H, H])$. But $[H, H]$ is also abelian, and we have a central series $H \rhd [H, H] \rhd \{e\}$, so that $H$ is nilpotent of nilpotency class 2.

In GAP, you can display the nilpotency class of some groups:

```
gap> NilpotencyClassOfGroup(DihedralGroup(4));
1
gap> NilpotencyClassOfGroup(DihedralGroup(8));
2
```

The command `DihedralGroup(2n)` returns the group $D_n$. The group $D_2$ is abelian and therefore of nilpotency class 1. The group $D_4$ has its center as commutator subgroup, $D_4' = C(D_4) = \{id, d\}$ where $d$ is the rotation by 180°. Thus, $D_4$ has nilpotency class 2.

Whether a group is nilpotent at all can be seen as follows:

```
gap> IsNilpotentGroup(DihedralGroup(8));
true
gap> IsNilpotentGroup(DihedralGroup(10));
false
```

## Exercises

1. Prove that a subgroup of a solvable group is solvable.
2. Are the dihedral groups $D_n$ solvable?
3. Prove that if $H$, $J$ are solvable groups, then $G = H \times J$ is solvable.
4. Prove that subgroups of nilpotent groups are nilpotent.

# Chapter 9
# The Hyperbolic Plane

In this chapter, we provide an introduction to hyperbolic geometry. In the first section, we describe the axiomatic approach to geometry and introduce the hyperbolic plane using Poincaré's disk model. In the next section, we consider the isometries of the hyperbolic plane and gain an understanding of them. Finally, we examine the decompositions of the hyperbolic plane and find that the decompositions of type $(n, m)$ which are not spherical or Euclidean can be realized in the hyperbolic plane.

## 9.1  Axiomatic Geometry

Here we provide a very brief introduction to the axiomatic approach to geometry, covering only what we will need later. Very readable introductions to the geometric content of this chapter can be found, for example, in [FG91] and [Ced91].

Most of the geometric facts in this book so far have concerned Euclidean geometry. This is the geometry that one learns in school and that we developed in the first chapter. In Sect. 7.4, however, we also dealt with geometry on the 2-sphere. There, suddenly different rules apply: There are no parallel lines and no translations, but there are more regular decompositions than in Euclidean geometry (see Theorem 7.19).

So far, we have naively practiced geometry in the sense that we assumed that everyone knows the Euclidean plane and terms like 'line' and 'point' are familiar.

EUCLID summarized the geometric knowledge of his time about 2300 years ago in the *Elements*. He introduced five axioms, which he used as a basis to prove geometric theorems. His axioms are as follows:

*It shall be required that*

1. *one can draw a line from any point to any point,*
2. *one can continuously extend a finite line,*

© The Author(s), under exclusive license to Springer-Verlag GmbH, DE,
part of Springer Nature 2024
S. Rosebrock, *Visual Group Theory*, Springer Undergraduate Mathematics Series,
https://doi.org/10.1007/978-3-662-69365-0_9

3. *one can draw the corresponding circle for any center and radius,*
4. *all right angles are equal to each other,*
5. *if a line, when intersecting with two other lines, causes the angles formed on the same side to be less than two right angles, then the two lines meet on the side where the angles are less than two right angles.*

Euclid's approach, namely to draw conclusions from undefined terms like *point* and *line* and *axioms*, i.e., unprovable postulates, to prove theorems and practice geometry, is still current. However, Euclid's system of axioms is not complete. In 1930, HILBERT published a complete system of axioms for Euclidean geometry in his *Foundations of Geometry*.

We then worked in a *model of the Euclidean plane*, namely the usual Euclidean plane $\mathbb{R} \times \mathbb{R}$ with coordinates and distance measurement according to Pythagoras: For points $P_1 = (x_1, y_1)$ and $P_2 = (x_2, y_2)$ their distance is

$$d_e(P_1, P_2) = \sqrt{(x_1 - x_2)^2 + (y_1 - y_2)^2}. \tag{9.1}$$

A model of a geometry is therefore a realization of the undefined terms *point, line* and point is *incident with* line, so that the axioms of this geometry are valid.

A more convenient version of the 5th axiom for our purposes is due to Playfair (1795):

5.' *For a line and a point not on the line, there is exactly one line through the point that is parallel to the original line.*

Two lines are called *parallel* if they do not intersect. This axiom is violated on the 2-sphere, as we have already seen (see Sect. 7.5). Any two different lines (great circles) on the 2-sphere intersect at exactly 2 points. If we identify all pairs of diametrically opposite points on the 2-sphere, we obtain the *projective plane*. In it, any two different lines (i.e., lines that come from great circles of the 2-sphere) intersect at exactly one point. The two intersection points on the surface of the sphere are diametrically opposite points, which are identified as one. The associated geometry is called *elliptic geometry*, and the projective plane with the corresponding definition of lines serves as a model of elliptic geometry. In this geometry, the 5th axiom would be replaced by:

5e. *For a line and a point not on the line, every line through the point intersects the original line.*

However, in this chapter we want to deal with *hyperbolic geometry*. This geometry is obtained by replacing axiom 5 by the following:

5h. *For a line and a point not on the line, there are at least two different lines through the point that do not intersect the original line.*

The five axioms above are not sufficient for hyperbolic geometry, and for this geometry too, HILBERT provided a complete set of axioms. As in Euclidean geometry, there are lines and points in hyperbolic geometry. Strictly axiomatically,

**Fig. 9.1** The hyperbolic
plane with a line

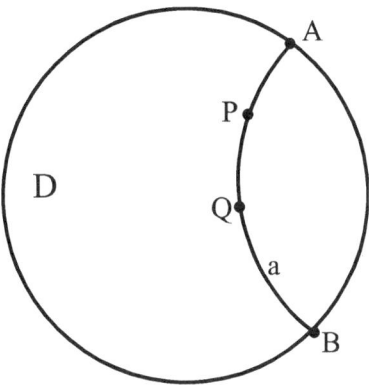

lines and points can be any two sets that satisfy the axioms. We now present the
*Poincaré model* of the hyperbolic plane, without naming all the axioms of hyperbolic
geometry or even showing that these axioms are all valid in this model. More
precisely, it is the Poincaré disk model of the hyperbolic plane, as Poincaré also
designed a half-plane model. For details on the Poincaré disk model, see [Moi90].

The *points* of the *hyperbolic plane* $\mathbb{H}^2$ are the interior points of a disk $D \subset \mathbb{R}^2$
with radius 1, but without the boundary points. The *lines* of the hyperbolic plane
are either diameters in $D$ (of course without the boundary points) or arcs of circles
that intersect the boundary at right angles. A typical hyperbolic line is the line $a$
in Fig. 9.1. When we speak of lines in this chapter, we always mean hyperbolic
lines. Euclidean lines are henceforth called *e-lines*. Points on the boundary of the
hyperbolic plane are not part of the hyperbolic plane. They are called *points at $\infty$*.

Angles are measured as in Euclidean geometry, i.e., the angle between two
hyperbolic lines is the Euclidean angle between their tangents at the intersection
point. However, distances are not measured as in Euclidean geometry according to
formula (9.1). If we have two different points $P, Q \in \mathbb{H}^2$, there is a line on which
the two points lie. In Fig. 9.1 the points $P, Q$ lie on the hyperbolic line $a$. $A, B$ are
the two points at $\infty$ for the line $a$. Let $e(P, Q)$ be the Euclidean distance on the
e-circle arc $a$. Then the *hyperbolic distance* between $P$ and $Q$ is defined as

$$d_h(P, Q) = \frac{1}{2} \left| \ln \frac{e(P, B) \cdot e(Q, A)}{e(P, A) \cdot e(Q, B)} \right|, \tag{9.2}$$

where ln is the natural logarithm. We observe two important properties: First, if you
swap $P$ and $Q$ in formula (9.2), the fraction changes to its reciprocal. The logarithm
has the opposite sign, which leaves the absolute value unchanged. It follows that

$$d_h(P, Q) = d_h(Q, P)$$

which must be fulfilled by any distance function (see Definition 4.53). Second,
if you let the point $P$ approach the point $A$ on $a$, then $e(P, A)$ approaches 0

and thus $d_h(P, Q)$ approaches infinity. Lines in hyperbolic geometry are therefore infinitely long, and the closer you get to the boundary, the greater distances become to overcome the corresponding Euclidean distance. If you want to go from one point near the boundary to another point near the boundary, the shortest path will initially lead towards the middle, in a circular arc, because distances are shorter there. Just like in Euclidean and elliptical geometry, *lines* are the realizations of shortest connections.

Let $\mathcal{H}$ be the group of isometries of the hyperbolic plane, i.e., the distance-preserving mappings from $\mathbb{H}^2$ onto itself, where lengths are measured with $d_h$. Two figures in the hyperbolic (Euclidean) plane are called *equivalent* if there is an isometry $g \in \mathcal{H}$ ($g \in \mathcal{E}$) that maps one figure onto the other. A property of a figure can then be identified with the set of all figures that have this property. A property is called *geometric* if every equivalent figure has the same property. A property is then a union of orbits of figures under the operation of the group $\mathcal{H}$ (the group $\mathcal{E}$).

Now one can define *hyperbolic geometry* as the properties of the hyperbolic plane that do not change under the group $\mathcal{H}$. Similarly, *Euclidean geometry* is the set of properties that do not change under the group $\mathcal{E}$. A geometry therefore includes a space and a group that operates on this space. This view of geometry as an operation of groups was first formulated by FELIX KLEIN in 1872 in his famous *Erlanger Program*.

## 9.2   Isometries in the Hyperbolic Plane

What isometries are possible in the hyperbolic plane? We start with the description of a *reflection*. For this, we consider a line $a \in \mathbb{H}^2$, over which we want to reflect (see Fig. 9.2). The hyperbolic line $a$ is part of a Euclidean circle $K$ with center $S$ and e-radius $r$. Now let $P \in \mathbb{H}^2$ be a point that we want to reflect over $a$. The image point $Q$ under this reflection lies on the e-line through $P$ and $S$ (dashed in Fig. 9.2). Let $e(P, S)$ be the Euclidean distance from $P$ to $S$. Then $Q$ is chosen so that

$$e(P, S) \cdot e(Q, S) = r^2. \tag{9.3}$$

If the line $a$ is a diameter of $D$, the reflection is just the Euclidean reflection. If $P \in a$, then Eq. (9.3) becomes $r \cdot r = r^2$, and the image point is again $P$ and thus also lies on $a$. The mirror line therefore remains fixed under the reflection, as is usual with reflections. The boundary circle of points at infinity is mapped onto itself. It is not difficult to see that, under this mapping, points of $\mathbb{H}^2$ are mapped to points of $\mathbb{H}^2$. In doing so, the points from one side of the line $a$ are mapped to the points on the other side and vice versa.

This mapping can be performed for the entire Euclidean plane given a circle $K$, except for the center point $S$. It is called *inversion* on the circle.

Let $M$ be the set of *e*-lines and the set of *e*-circles in the Euclidean plane. It can be shown that an inversion on the circle maps elements of $M$ onto elements of $M$.

**Fig. 9.2** Reflection on a
hyperbolic line

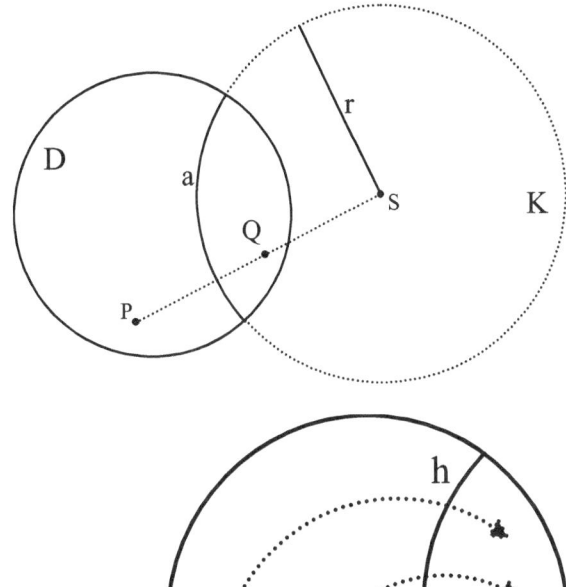

**Fig. 9.3** Rotation about a
point at ∞

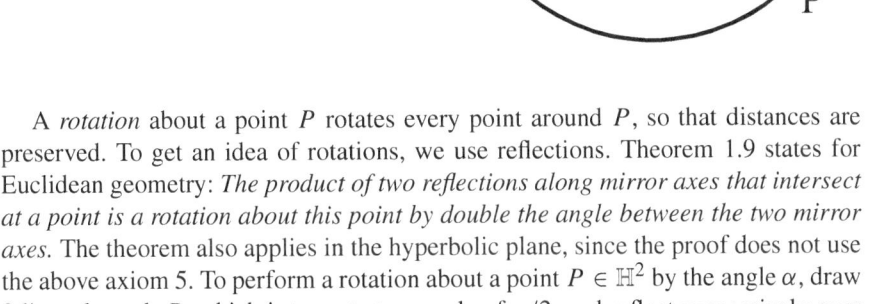

A *rotation* about a point $P$ rotates every point around $P$, so that distances are preserved. To get an idea of rotations, we use reflections. Theorem 1.9 states for Euclidean geometry: *The product of two reflections along mirror axes that intersect at a point is a rotation about this point by double the angle between the two mirror axes*. The theorem also applies in the hyperbolic plane, since the proof does not use the above axiom 5. To perform a rotation about a point $P \in \mathbb{H}^2$ by the angle $\alpha$, draw 2 lines through $P$, which intersect at an angle of $\alpha/2$, and reflect successively over the two lines. The reader is encouraged to draw corresponding pictures.

In hyperbolic geometry, there are special forms of parallelism. Two parallel lines that have a common point at ∞ are called *asymptotic*. To make it clear again, the point at ∞ does not belong to the lines, nor does it belong to the hyperbolic plane. It is rather something like an "endpoint" or "limit point" of the lines. If parallel lines are not asymptotic, they are called *ultraparallel*.

If you reflect over two asymptotic lines in succession, you have something like a rotation around a point at infinity, a so-called *limit rotation*. Let $P$ be such a point of the two asymptotic lines $g$, $h$. If you first reflect over $g$ and then over $h$, you get a "rotation" as in Fig. 9.3. This rotation has infinite order in the group $\mathcal{H}$.

**Fig. 9.4** Translation through
two reflections

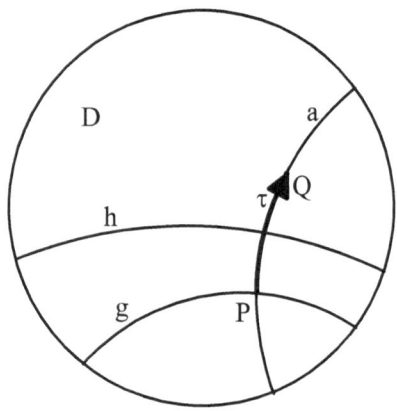

We use Theorem 1.9 once again to get an idea of *translations*. In Euclidean as
well as in hyperbolic geometry the following holds: *The product of two reflections
along parallel lines is a translation perpendicular to the mirror axes by twice their
distance.*

**Theorem 9.1** *For $P, Q \in \mathbb{H}^2$ there exists a translation $\tau \in \mathcal{H}$ with $\tau(P) = Q$, i.e.,
$\mathcal{H}$ is transitive on the set of points of $\mathbb{H}^2$.*

**Proof** Let $a \in \mathbb{H}^2$ be the line incident to $P$ and $Q$. Let $g \in \mathbb{H}^2$ be the line
perpendicular to $a$ through $P$ and $h \in \mathbb{H}^2$ be the line perpendicular to the
midpoint of the line from $P$ to $Q$ (see Fig. 9.4). $g$ is parallel to $h$ because both
are perpendicular to the same line $a$. Then $\tau = s_h s_g$ is a translation, where $s_g$ is the
reflection over $g$. It is easy to see that $\tau(P) = Q$ holds.                          □

In the Euclidean plane, there are infinitely many parallel axes for a translation.
An *axis* is a line that is mapped onto itself by a translation. In hyperbolic geometry,
there is only one axis, lines parallel to it are not mapped onto themselves by the
translation.

A *glide reflection* is, as in Euclidean geometry, the product of a reflection and a
translation along the mirror line.

It can be proven that lines are transformed into lines by isometries and angles do
not change their size under isometries. We can then conclude:

**Theorem 9.2** *Every triangle has an angle sum less than $\pi$.*

**Proof** According to Theorem 9.1 there is a translation that transforms one vertex
of the triangle into the center of $D$ (our model of the hyperbolic plane). As a result,
two sides of the triangle become Euclidean lines. These triangle sides are parts of
two diameters of $D$.

The angles on the third side are, since the side is a circular arc, smaller than the
angles one would get if the side was a Euclidean line instead of a circular arc. The
sum of the angles is therefore smaller than in the Euclidean plane and therefore
smaller than $\pi$.                                                                       □

**Fig. 9.5**   An asymptotic
triangle

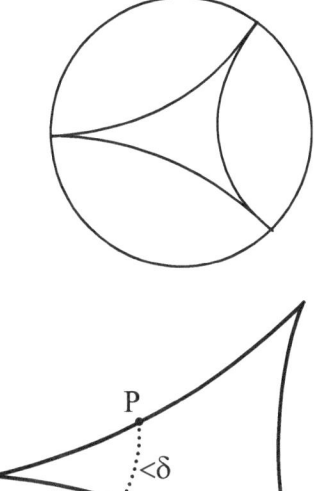

**Fig. 9.6**   A "thin" triangle in
the hyperbolic plane

The sum of the angles of a Euclidean triangle is exactly $\pi$, and the sum of the angles of a triangle on the 2-sphere is always greater than $\pi$. In hyperbolic geometry, there are also triangles where each angle is $0°$ (see Fig. 9.5). Such a triangle is called an *ideal triangle*. Strictly speaking, this is not a triangle at all, because the lines (asymptotically) run parallel and therefore have no intersection point. Any two such triangles can be transformed into each other by an isometry. Given two different ideal triangles with vertices $A, B, C$ and $A', B', C'$ (the vertices are read in clockwise order of appearance on the boundary). Then one can map point $A$ to $A'$ with a limit rotation. With another limit rotation, this time around point $A'$, one can map $B$ to $B'$. With a translation along the hyperbolic line $\overline{A'B'}$ $C$ can finally be mapped to $C'$, and the composition of the three mappings thus maps one triangle onto the other. So there is only one ideal triangle up to isometry.

In hyperbolic geometry, triangles are thin in the following sense: Let $a, b, c \in \mathbb{H}^2$ be sides of a triangle. Then there is a constant $\delta \in \mathbb{R}$ such that every point $P$ on $a$ has a distance of at most $\delta$ to $b$ or to $c$. The triangle is then called $\delta$-thin (see Fig. 9.6).

**Theorem 9.3** *Every triangle in the hyperbolic plane is $\delta$-thin, where $\delta = \ln(1 + \sqrt{2})$.*

***Proof*** It is easy to see that if a triangle in the hyperbolic plane is $\delta$-thin with the smallest possible $\delta$, then the diameter $d$ of its in-circle satisfies $d \geq \delta$.

The triangle with the largest in-circle is the asymptotic triangle of Fig. 9.5. To this end, let an arbitrary, non-asymptotic triangle with vertices $A, B, C$ be given. Moving the vertex $A$ along the straight line $AB$ towards infinity enlarges the in-circle. The in-circles become larger and are contained within each other. If a similar process is carried out with the other two vertices, an asymptotic triangle with a larger in-circle than the original triangle is finally obtained.

Finally, one convinces oneself that the in-circle diameter for asymptotic triangles is exactly $\ln(1 + \sqrt{2})$. The calculation is purely technical, and we omit it.   □

For the Euclidean plane, it is easy to see that there is no constant $\delta$ such that every triangle would be $\delta$-thin, since one can make a triangle arbitrarily large. For example, an equilateral triangle with edge length $4\delta$ in the Euclidean plane has a distance greater than $\delta$ from a midpoint of an edge to the other two edges.

Every isometry is, as in the Euclidean plane, determined by three points $P$, $Q$, $R$, which do not lie on a straight line (see Corollary 1.8 on page 9). A point $A$ is uniquely determined by its distances to $P$, $Q$ and $R$ and thus also its image point.

**Theorem 9.4**  *Every isometry of the hyperbolic plane is the product of at most three reflections. In particular, $\mathcal{H}$ is generated by reflections.*

**Proof**  Let $g \in \mathcal{H}$ be an arbitrary isometry, and $P$, $Q$, $R \in \mathbb{H}^2$ be three points that do not lie on a straight line. With a reflection $s_a$ over the perpendicular bisector $a$ of the line segment from $P$ to $g(P)$, the point $P$ is already correctly mapped. If $s_a(Q) \neq g(Q)$, then we reflect over the line $b$ through $g(P)$ and the midpoint of the line segment from $s_a(Q)$ to $g(Q)$. Thus, $P$ and $Q$ are already correctly mapped. With at most one more reflection over the line through $g(P)$ and $g(Q)$, $s_b s_a(R)$ is mapped to $g(R)$.   □

It can be proven (see for example [Sti92]) that every isometry of the hyperbolic plane is either a rotation, a reflection, a limit rotation, a translation or a glide reflection. However, the orientation-preserving isometries of the hyperbolic plane have a different classification according to modern terminology (see also [Lö17]). There are exactly the following three types of orientation-preserving isometries:

- *hyperbolic*: A non-trivial orientation-preserving isometry of the hyperbolic plane is called *hyperbolic* if it has an axis but no fixed points. A hyperbolic isometry corresponds to a translation.
- *parabolic*: A non-trivial orientation-preserving isometry of the hyperbolic plane is called *parabolic* if it has no fixed points and no axis. A parabolic isometry corresponds to a limit rotation.
- *elliptic*: A non-trivial orientation-preserving isometry of the hyperbolic plane is called *elliptic* if it has a fixed point. An elliptic isometry corresponds to a rotation.

## Exercises

1. Show that every element of $\mathcal{H}$ maps lines to lines.
2. Prove that $\mathcal{H}$ is transitive on the set of lines of the hyperbolic plane.
3. In Euclidean geometry, lines are parallel if and only if they are *equidistant*, i.e., they have at every point the same distance from each other. Does the same apply in hyperbolic geometry?

## 9.3   Decompositions of the Hyperbolic Plane

Two figures in the hyperbolic plane are called *congruent* if there is an isometry in
$\mathcal{H}$ that maps one figure onto the other.

**Theorem 9.5** *Let $\alpha, \beta, \gamma > 0$ be angles such that $\alpha + \beta + \gamma < \pi$. Then there is a
triangle in the hyperbolic plane with the interior angles $\alpha, \beta, \gamma$ (in cyclic order in
the mathematically positive sense), and any two such triangles are congruent.*

**Proof** We will construct a triangle where one vertex, namely the vertex $A$ with the
angle $\alpha$, is exactly in the center of the disc $D$ of our model of the hyperbolic plane.
This point does not differ from the other points of the hyperbolic plane, just as the
origin of a coordinate system of the Euclidean plane. With a different coordinate
system, it would be elsewhere.

   We start with an arbitrary e-circle $K$ with center $S$ in the Euclidean plane. Let
$\delta = \pi - (\alpha + \beta + \gamma)$. Then $0 < \delta < \pi$. Let $B, C$ be points on the circle $K$, such
that the e-lines $B, S$ and $C, S$ form the angle $\delta$ at the point $S$ (see Fig. 9.7). Let $b$
and $c$ be e-rays through $B$ and $C$, which form the angles $\beta$ and $\gamma$ with the shorter
segment of $K$ between $B$ and $C$. By definition of $\delta$, we have

$$\delta + \beta + \gamma = \pi - \alpha < \pi.$$

So $\delta + \beta + \gamma + \pi < 2\pi$, which means that an e-quadrilateral is formed, i.e., the
e-rays $b, c$ must intersect. The size of the angle at the intersection point $A$ is $\alpha$,
because $\delta + \beta + \gamma + \alpha + \pi = 2\pi$, the sum of angles in the e-quadrilateral.

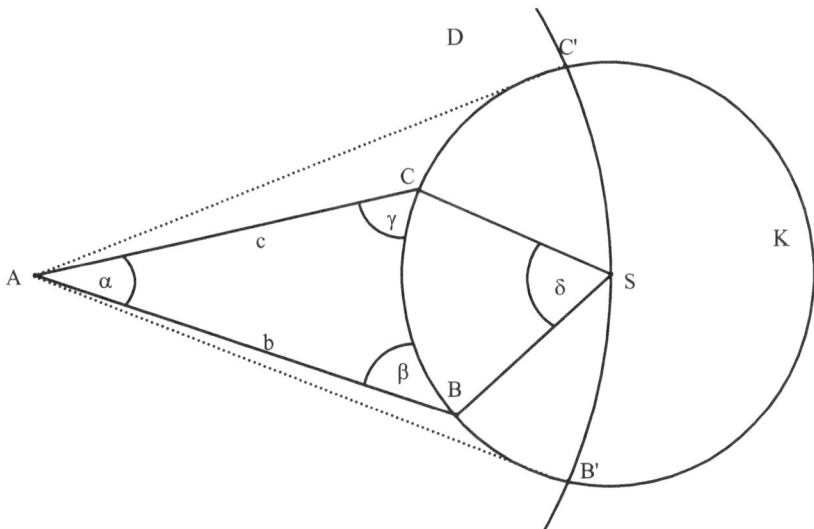

**Fig. 9.7**  Construction of a triangle with given angles

Let $D$ be the disc with center $A$, whose boundary intersects the circle $K$ at right angles. $D$ is the model of our hyperbolic plane. The boundary of $D$ intersects $K$ at the points $B'$ and $C'$. In $D$, the triangle $A, B, C$ is a hyperbolic triangle with the required angles.

The points $B$ and $C$ are inside the disc $D$. The e-lines $c$ and $b$ meet $K$ at the angles $\gamma$ and $\beta$, while the e-lines $A, C'$ and $A, B'$ are tangents to $K$. Therefore the points $B$ and $C$ must be closer to $A$ than the points $B'$ and $C'$.

From the construction it follows that the triangle is unique, except for a rotation about the point $A$. Therefore, all other triangles with the same interior angles are congruent to the constructed one.                                                 □

Hyperbolic geometry is clearly different from Euclidean Geometry. In Euclidean geometry, triangles with the same angles are generally not congruent, but only similar, i.e., they can be transformed into each other by a central dilation.

Furthermore, the theorem remains true (with only a slightly modified proof), if one or more of the interior angles are 0, i.e., if the vertices lie on the boundary of the hyperbolic plane. So there is up to congruence only one triangle like the one in Fig. 9.5, where all vertices lie on the boundary.

In the following, we are looking for decompositions of the hyperbolic plane of type $(n, m)$, i.e., we are looking for decompositions of $\mathbb{H}^2$ into regular $n$-gons, of which always $m$ meet at one vertex. The regular decompositions of the Euclidean plane are of type $(3, 6)$, $(4, 4)$ and $(6, 3)$. We have dealt with the decomposition of type $(3, 6)$ in detail in Sect. 4.6, and the other two types appeared in the exercises of Sect. 4.6. See also Exercise 1. of Sect. 6.1. We have investigated the regular decompositions of the 2-sphere in Theorem 7.19 on page 141. All other regular pairs of positive integers belong to decompositions of the hyperbolic plane:

**Theorem 9.6** *There are infinitely many different regular decompositions of the hyperbolic plane.*

**Proof** Let there be a regular decomposition of type $(n, m)$ of the hyperbolic plane. Due to Theorem 9.2, the sum of angles in each $n$-gon is less than $(n - 2)\pi$. Since all interior angles in a regular $n$-gon are equal, each angle is less than $(1 - 2/n)\pi$.

At each vertex, $m$ of these angles come together, which must total $2\pi$. So it follows that

$$m(1 - 2/n)\pi > 2\pi.$$

This is equivalent to

$$\frac{1}{m} + \frac{1}{n} < \frac{1}{2}. \tag{9.4}$$

Every pair of numbers $(n, m)$ that satisfies this inequality leads to a decomposition of the hyperbolic plane. To see this, we construct a triangle $d \in \mathbb{H}^2$ with the interior

**Fig. 9.8** Fundamental domain with regular hexagon

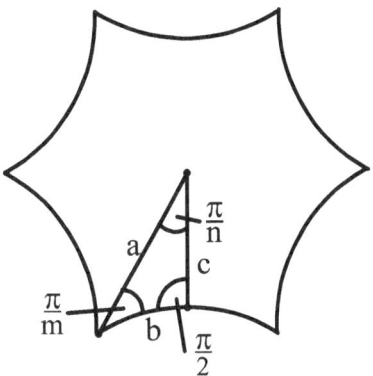

angles $\pi/n$, $\pi/m$, $\pi/2$. The sum of these three angles is, due to (9.4), less than $\pi$, and then Theorem 9.5 gives us the existence of such a triangle.

The triangle $d$ becomes the fundamental domain of the decomposition of type $(n, m)$. If $a$ and $c$ are the edges of the triangle that enclose the angle $\pi/n$, the reflections over $a$ and $c$ generate group elements which map the fundamental domain $d$ onto a regular $n$-gon (see Fig. 9.8). The group elements are $1, a, ac, aca, (ac)^2, (ac)^2a, (ac)^3, \ldots, (ac)^{n-1}, (ac)^{n-1}a$, as can already be traced in the case of the Euclidean plane in Sect. 4.6.

Adding the generator $b$, you can reflect the regular $n$-gon, analogous to the Euclidean case, as far as you like and get the desired decomposition. □

Analogous to the Euclidean case in Sect. 4.6, one shows:

**Theorem 9.7** *If $\frac{1}{m} + \frac{1}{n} < \frac{1}{2}$, then the group of the decomposition of the hyperbolic plane of type $(n,m)$ is presented by*

$$G_{(n,m)} = \langle s_a, s_b, s_c \mid s_a^2, s_b^2, s_c^2, (s_a s_c)^n, (s_a s_b)^m, (s_c s_b)^2 \rangle.$$

**Example 9.8** We consider the decomposition of the hyperbolic plane of Fig. 9.9 into regular pentagons, of which 4 meet at each vertex.

$1/4 + 1/5 < 1/2$, and so this is an example of Theorem 9.6. The associated group is presented by

$$G_{(5,4)} = \langle s_a, s_b, s_c \mid s_a^2, s_b^2, s_c^2, (s_a s_c)^4, (s_a s_b)^5, (s_c s_b)^2 \rangle$$

and is infinite.

One can further generalize Theorem 9.7 by considering three arbitrary natural numbers $p, q, r$, with $1/p + 1/q + 1/r < 1$. Then $\pi/p + \pi/q + \pi/r < \pi$, and due to Theorem 9.5, there is a triangle with interior angles $\pi/p, \pi/q, \pi/r$. Using this triangle as a fundamental domain, one obtains a decomposition of the

**Fig. 9.9**   Decomposition of
the hyperbolic plane

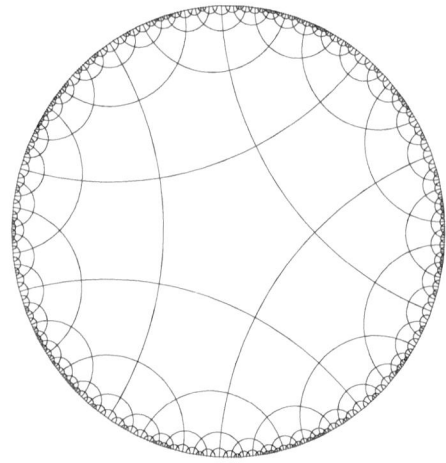

**Fig. 9.10**   Decomposition of
the hyperbolic plane

hyperbolic plane and as the symmetry group of this decomposition, the following
*triangle group*:

$$\langle s_a, s_b, s_c \mid s_a^2, s_b^2, s_c^2, (s_a s_c)^p, (s_a s_b)^q, (s_c s_b)^r \rangle.$$

Let $a, b, c$ be the three lines of Fig. 9.5. They form a "triangle" $d$, which can be
taken as a fundamental domain. Then, for the group generated by the reflections
over the three lines, we have

$$\langle s_a, s_b, s_c \mid s_a^2, s_b^2, s_c^2 \rangle = \mathbb{Z}_2 * \mathbb{Z}_2 * \mathbb{Z}_2,$$

because the product of any two of these reflections is a limit rotation and this has
infinite order in $\mathcal{H}$. In Fig. 9.10, all reflections of this group are depicted.

The group of a decomposition is an example of a *Coxeter group* (see [Rat94]). These are discontinuous groups generated by finitely many reflections $s_i$ on *hyperplanes* (i.e., $n-1$-dimensional subspaces) in $n$-dimensional Euclidean or other space. If the product of two such reflections $s_i s_j$ has finite order, it must be a rotation by an angle $2\pi/m_{ij}$ for a natural number $m_{ij}$. The presentation of such a Coxeter group then has generators $s_i$ and defining relations $(s_i s_j)^{m_{ij}}$, where $m_{ij} = m_{ji}$ and all $m_{ii} = 1$. If the product $s_i s_j$ has infinite order, i.e., the corresponding hyperplanes do not intersect, one sets $m_{ij} = \infty$, and the relation $(s_i s_j)^{m_{ij}}$ does not appear in the presentation. The simplest example is the group $\langle s_1 \mid s_1^2 \rangle$ generated by a reflection in the Euclidean plane. But also

$$D_n = \langle s_1, s_2 \mid s_1^2, s_2^2, (s_1 s_2)^n \rangle$$

is a Coxeter group. Likewise, the groups $G_{(n,m)}$ in our given presentations are Coxeter groups. Also, the groups $S_n$ with the presentations of Theorem 5.19 are examples of Coxeter groups.

## Exercises

1. Provide group presentations for decompositions of the hyperbolic plane that come from a fundamental domain, which is a triangle, with exactly one (exactly two) point(s) on the boundary of the hyperbolic plane.
2. Consider the decomposition of type (5,4) of the hyperbolic plane. Let $f$ be a regular pentagon of this decomposition. The lines on the boundary of this pentagon are $a, b, c, d, e$, read in order. Show that

   (a) the group generated by $s_a, s_b, s_c, s_d, s_e$, with $f$ as a fundamental domain, has the presentation

   $$G = \langle s_a, s_b, s_c, s_d, s_e \mid$$

   $$s_a^2, s_b^2, s_c^2, s_d^2, s_e^2, (s_a s_b)^2, (s_b s_c)^2, (s_c s_d)^2, (s_d s_e)^2, (s_e s_a)^2 \rangle,$$

   (b) the subgroup of orientation-preserving isometries has the presentation

   $$G^+ = \langle d_1, d_2, d_3, d_4, d_5 \mid d_1^2, d_2^2, d_3^2, d_4^2, d_5^2, d_1 d_2 d_3 d_4 d_5 \rangle,$$

   where $d_1 = s_a s_b, d_2 = s_b s_c, \ldots, d_5 = s_e s_a$. The fundamental domain then consists of $f$ together with any 5-gon adjacent to $f$.

3. Write the symmetry group of the rhombus as a presentation of a Coxeter group.

# Chapter 10
# Hyperbolic Groups

In this chapter, groups are studied as geometric objects. The Cayley graph of a group fully describes the group. Through the operation of a group on its Cayley graph of Theorem 4.52, we have an operation of a group on a geometric object.

In the first section, a method is described to represent a relation of a given presentation as a graph in the plane. In the following section, a concept of isometry for groups is defined, quasi-isometry. After that, methods for solving the word problem in groups are described. In the section on hyperbolic groups, curvature phenomena in groups are considered. They allow a particularly simple solution to the word problem. In the last section on combings, the concept of the hyperbolic group is generalized, but only to the extent that the word problem can still be solved.

## 10.1 Van Kampen Diagrams

A word $w$ in the generators $X$ and their inverses of a presentation $P = \langle X \mid R \rangle$ is, according to Theorem 5.2, a relation if and only if the corresponding path, starting from the vertex 1 in the Cayley graph, which reads the word $w$, is closed. Theorem 5.14 describes words that are relations, algebraically: They are conjugate products of defining relations. Every relation in $P$ can be written as a finite product $\prod_{i_j} w_j r_{i_j}^{\epsilon_j} w_j^{-1}$, where $\epsilon_j = \pm 1, r_{i_j} \in R$ and $w_j \in F(X)$. Every word in $X$ which can be represented in this way is conversely a relation in $P$. This provides the possibility to represent relations geometrically by assembling defining relations, taking an $n$-gon, a *disk*, for each defining relation of length $n$. This geometric representation of relations is described in this section.

A graph is called *planar* if it can be drawn in the plane without crossing edges.

**Definition 10.1** Let $P = \langle X \mid R \rangle$ be a group presentation and $w$ a relation in $P$. Let $\Gamma$ be a finite, connected, planar, oriented graph, where each edge is labeled with

S. Rosebrock, *Visual Group Theory*, Springer Undergraduate Mathematics Series, https://doi.org/10.1007/978-3-662-69365-0_10

**Fig. 10.1**   van Kampen
diagram for $b^2a^2b^{-2}a^{-2}$

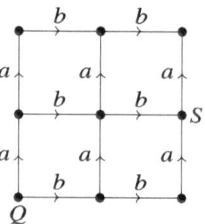

a generator, so that the boundary of each $n$-gon reads a defining relation. In the
boundary of $\Gamma$ lies the *base point* $Q$, so that when you read the boundary of the
area determined by $\Gamma$ counterclockwise starting at $Q$, you get the word $w$. Then $\Gamma$
is called a *van Kampen diagram* for $w$.

**Example 10.2**   Let $P = \langle a, b \mid ab = ba \rangle$. Figure 10.1 shows a van Kampen
diagram for $b^2a^2b^{-2}a^{-2}$.

If you start at the bottom left in Fig. 10.1, at the base point $Q$, and read the
diagram counterclockwise, you read exactly $b^2a^2b^{-2}a^{-2}$. The diagram is also a
Van Kampen diagram for the word $ab^{-2}a^{-2}b^2a$, where we start to read from the
base point $S$.

Van Kampen diagrams are motivated by the following theorem:

**Theorem 10.3**   *Let $P = \langle X \mid R \rangle$ be a group and $v$ a word in the generators and
their inverses. Then $v$ is a relation in $P$ if and only if there is a van Kampen diagram
for $v$.*

**Proof**   If $v$ is a relation, then according to Theorem 5.14, $v$ can be written as a
product of conjugates of defining relations, i.e.:

$$v = \prod_{j=1}^{n} w_j r_{i_j}^{\epsilon_j} w_j^{-1},$$

where $\epsilon_j = \pm 1$, $r_{i_j} \in R$ and $w_j \in F(X)$. Figure 10.2 shows the corresponding van
Kampen diagram. Each circle reads one of the relations $r_{i_j}$ on its boundary. Start at
the point where the corresponding $w_j$ ends, the *base point* of the disk. Depending on
$\epsilon_j$, the respective relation is read clockwise or counterclockwise. If you read around
this diagram counterclockwise, you read exactly $v$.

Conversely, let a van Kampen diagram $\Gamma$ for $v$ with base point $Q$ be given. $Q$ is
the vertex from which you read $v$ around the edge of $\Gamma$. Each disk $d$ also has a base
point on its boundary: the vertex at which you have to start reading the boundary of
$d$ to read the associated relation.

Enumerate the disks of $\Gamma$ and choose for the disk with number $j$ a path $w_j$ from
$Q$ to the base point of the disk $j$. Cut along $w_j$. Instead of formally describing this
process, we perform it using the example of Fig. 10.1 (see Fig. 10.3). The disk $j$

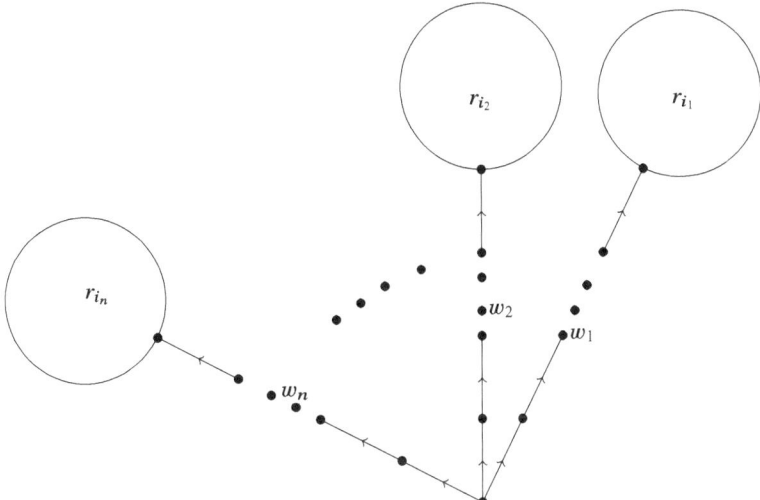

**Fig. 10.2** van Kampen diagram for $v$

**Fig. 10.3** Cut open van Kampen diagram of Fig. 10.1

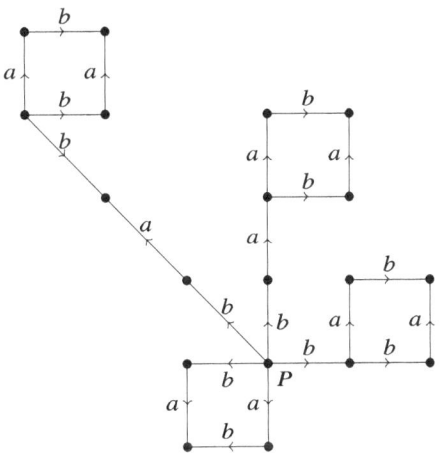

reads the relation $r_{i_j}$. Let $\epsilon_j$ be $+1$ if the relation is read counterclockwise and $-1$ otherwise. Then $v$ can be written as a product of conjugates:

$$v = \prod_{j=1}^{n} w_j r_{i_j}^{\epsilon_j} w_j^{-1},$$

where $\epsilon_j = \pm 1$, $r_{i_j} \in R$ and $w_j \in F(X)$. But if a word can be expressed as a product of conjugates of defining relations, then according to Theorem 5.14 it is a relation. $\qquad\square$

We write the word $v = b^2a^2b^{-2}a^{-2}$ of Example 10.2 as a product of conjugates and read Fig. 10.3 for this:

$$v = b\ bab^{-1}a^{-1}\ b^{-1}\cdot ba\ bab^{-1}a^{-1}\ (ba)^{-1}\cdot bab^{-1}\ bab^{-1}a^{-1}\ ba^{-1}b^{-1}\cdot bab^{-1}a^{-1}.$$

The reader should check that this word reduces to $b^2a^2b^{-2}a^{-2}$.

**Example 10.4**  We introduced the quaternion group in Exercise 6. of Sect. 5.1. A presentation of this group is:

$$Q = \langle k, j \mid k^2 = j^2, j = kjk \rangle.$$

$k^4$ is a relation in $Q$, which is proven by the van Kampen diagram of Fig. 10.4.

**Definition 10.5**  The *valence* of a disk in a van Kampen diagram is the number of its boundary edges. If a disk touches itself at an edge, this edge counts twice. The *valence* of a vertex in a van Kampen diagram is the number of edges incident to it. An edge counts twice if both of its boundary vertices are the same vertex.

Given any van Kampen diagram, one can omit the vertices of valence 2 by writing not generators, but words on the edges: If one merges two edges that share a vertex of valence 2 into one, and if the two edges have the label $a$ and $b$, then afterwards the new edge will have the label $a^{\pm1}b^{\pm1}$ with signs depending on whether the directions of the original edges matches the direction of the merged edge.

In Fig. 10.5, the van Kampen diagram of Fig. 10.4 is shown again, but without the vertex of valence two incident with the edges labeled $j$ and $k$.

**Fig. 10.4**  van Kampen diagram for $k^4$

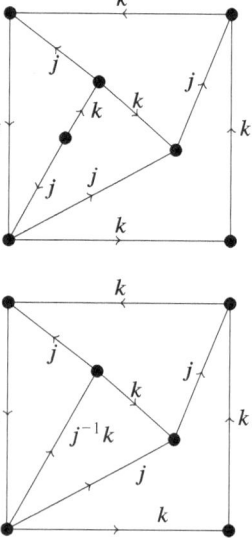

**Fig. 10.5**  van Kampen diagram with two edges merged

## Exercises

1. Provide a van Kampen diagram for the relation $d^2sd^2s$ in the group

$$D_3 = \langle s, d \mid s^2, d^3, sdsd \rangle.$$

If you have problems, read Sect. 5.1 again.
2. Prove that the presentation $Q$ of Example 10.4 is a presentation of the quaternion group by tracing back the relations of Exercise 6. of Sect. 5.1 to $Q$. Show that in $Q$ the relation $k = jkj$ holds.
3. Let $P = \langle a, b \mid a^3 = b^2, ab = ba^2 \rangle$. Draw a van Kampen diagram for the relation $a^3 = 1$.

## 10.2   Quasi-Isometries and the Švarc–Milnor Theorem

In Sect. 4.6, we saw that the Cayley graph of a group which operates on a decomposition is "almost isometric" to the decomposition. This fact is clarified by the ŠVARC–MILNOR theorem. We give here the exact formulation of this theorem, as described in [NS96]. A proof is in [dlH00], see also [BH99] and [Mil68]. The theorem was proven in the 1950s in Russia and rediscovered by MILNOR later.

We start with the definition of a quasi-isometry, which generalizes the concept of isometry. Here is the idea:

Two spaces are called quasi-isometric if, from a great distance, they look the same. For example, a line is not quasi-isometric to the plane, because, even from a very great distance, a line in the plane does not look like the plane. However, if you draw parallels to the y-axis through all integer points of the x-axis in the plane, then the resulting set of lines is quasi-isometric to the plane. From a great distance, the set of lines looks like the plane itself.

Let $(X, d_X)$ and $(Y, d_Y)$ be metric spaces. A mapping $f : (X, d_X) \rightarrow (Y, d_Y)$ is called a *quasi-isometric mapping* if there are numbers $\lambda, \epsilon \in \mathbb{R}$ such that $\forall a, b \in X$ the following holds:

$$\frac{1}{\lambda} d_X(a, b) - \epsilon \leq d_Y(f(a), f(b)) \leq \lambda d_X(a, b) + \epsilon.$$

A quasi-isometric mapping $f : X \rightarrow Y$ is a *quasi-isometry* between $X$ and $Y$ if there is a quasi-isometric mapping $g : Y \rightarrow X$ and a constant $k \in \mathbb{Z}$ such that $\forall x \in X \ d_X(x, g(f(x))) \leq k$ and $\forall y \in Y \ d_Y(y, f(g(y))) \leq k$. In this case, $X$ and $Y$ are called *quasi-isometric*.

The identical mapping of a metric space onto itself is a quasi-isometry. It is easy to see that quasi-isometry between metric spaces is an equivalence relation.

**Example 10.6** The map $f: \mathbb{Z}^2 \to \mathbb{R}^2$ where both spaces are equipped with the euclidean metric is a quasi-isometry.

The map $f: \mathbb{R} \to \mathbb{R}$ given by $f(x) = x^2$ is not a quasi-isometry.

**Theorem 10.7** *The Cayley graphs of two different finite generating systems of a group are quasi-isometric.*

**Proof** Let $(g_1, \ldots, g_n)$ and $(h_1, \ldots, h_m)$ be generating systems of the same group $G$. Let $\Gamma$ be the Cayley graph of $G$ with respect to the generators $(g_1, \ldots, g_n)$ and $\Gamma'$ the Cayley graph of $G$ with respect to the generators $(h_1, \ldots, h_m)$.

We can write each element $g_i$ in the generators $(h_1, \ldots, h_m)$ and also each $h_j$ in the generators $(g_1, \ldots, g_n)$. Let $c$ be the maximum length of the words that appear in this process.

The mapping $f: \Gamma \to \Gamma'$, which is the identity on the elements of $G$ but transforms words in the $g_i$ into words in the $h_j$, is a quasi-isometric mapping, where we can choose the numbers $\lambda$ and $\epsilon$ from the definition of quasi-isometric mapping $\lambda = c$ and $\epsilon = 0$. An edge in $\Gamma$ is replaced by at most $c$ edges in $\Gamma'$.

The same applies to a correspondingly constructed quasi-isometric mapping $g: \Gamma' \to \Gamma$. $f$ and $g$ are quasi-isometries, and the constant $k$ defined in the definition of quasi-isometry can be chosen as $(c^2 + 1)/2$.                                   $\square$

Conversely, one defines:
Two finitely generated groups are called *quasi-isometric* if they have Cayley graphs that are quasi-isometric. So, for example, any two finite groups are quasi-isometric to each other, because finite Cayley graphs are quasi-isometric. One can take as constant $\lambda$ the maximum distance between two points of the Cayley graph. From a great distance, every finite group looks like the trivial group.

Let $(X, d_X)$ be a *geodesic metric space*, i.e., shortest distances in $X$ can be realized by *geodesics*. This means, if $x, y \in X$, there is a path (a geodesic) in $X$ whose length is equal to $d_X(x, y)$ and which connects $x$ with $y$. For example, the Euclidean plane without the origin is not a geodesic metric space, because the point $-1$ on the $x$-axis cannot be connected with the point 1 on the $x$-axis by a path of length 2. The Cayley graph of a finitely generated group is a geodesic metric space, because the distance between two points can always be realized by a path.

An $\epsilon$-*neighborhood* $U_\epsilon(x)$ of a point $x \in X$ is defined as:

$$U_\epsilon(x) = \{y \in X \mid d_X(x, y) < \epsilon\}.$$

Let $G$ be a finitely generated group that operates on $X$ by isometries. This operation is discrete with compact fundamental domain. *Discrete* means that accumulation points are not allowed, so that

$$\forall x \in X, \exists \epsilon > 0 \text{ with } U_\epsilon(x) \cap g \cdot U_\epsilon(x) = \emptyset, \forall g \in G - \{1\}.$$

Let $(g_1, \ldots, g_n)$ be a generating system of $G$, and $P \in X$ be a fixed point. We define a mapping $\Phi \colon \Gamma_G(g_1, \ldots, g_n) \rightarrow X$. If $h \in G$ is a vertex of the Cayley graph, then $\Phi(h) = h(P)$. For each generator $g_i$ we also choose a path $p_i$ from $P$ to $g_i(P)$. An edge from $h' \in \Gamma_G$ to $h \in \Gamma_G$, labeled with $g_i$, is then mapped by $\Phi$ to $h'(p_i)$.

Then the following theorem of ŠVARC–MILNOR holds:

**Theorem 10.8** *Let* $G = \langle g_1, \ldots, g_n \rangle$ *be a finitely generated group that operates by isometries discretely on a non-empty geodesic metric space* $X$ *with a compact fundamental domain. Then the mapping* $\Phi \colon \Gamma_G(g_1, \ldots, g_n) \rightarrow X$ *is a quasi-isometry.*

All spaces on which a given group "nicely" operates are thus quasi-isometric. A quasi-isometry between a Cayley graph $\Gamma_G$ and a decomposition of the Euclidean plane, on which the associated group $G$ operates, is indicated in Sect. 4.6, where the Cayley graph was embedded into the decomposition.

## Exercises

1. Prove that the constant $k$ that appears in the proof of Theorem 10.7 can be chosen to be $(c^2 + 1)/2$.
2. Show that the Cayley graphs $\Gamma_{\mathbb{Z}}(1)$ and $\Gamma_{\mathbb{Z}}(2, 3)$ are quasi-isometric. Determine the occurring constants $\lambda, \epsilon$.

## 10.3 Isoperimetric Inequalities

Theorem 10.3 gives us in some cases a tool for solving the word problem in groups. Suppose we are given a word $w$ of length $n$ and we want to know whether or not it is trivial in a group given by a presentation. If a function $f \colon \mathbb{N} \rightarrow \mathbb{N}$ can be calculated such that every van Kampen diagram for a word of length $n$ needs at most $f(n)$ many disks, and if one knows $f(n)$, then one can build all possible, finitely many van Kampen diagrams that have at most $f(n)$ disks and whose boundary has length $n$. If there is one with boundary word $w$ among them, then one has proven that $w$ is a relation. If none of these van Kampen diagrams reads $w$ on its boundary, then $w$ is not a relation. This motivates the following definition.

We write $l(w)$ for the length of a word $w$ in the generators of a presentation and $A(w)$ for the minimum number of disks over all van Kampen diagrams for $w$ if $w$ is trivial in the corresponding group.

**Definition 10.9** Let $P = \langle X \mid R \rangle$ be a finite presentation of the group $G$. If a function $f : \mathbb{N} \to \mathbb{N}$ satisfies the *isoperimetric inequality*

$$f(n) \geq \max\{A(w) \mid l(w) \leq n, \ w \text{ trivial in } G\},$$

then $f$ is called an *isoperimetric function* of $P$.

We restrict ourselves in this definition to finite presentations, because otherwise one could take all relations as the set of relations and have the uninteresting isoperimetric function $f(n) = 1$ for every group.

Let $f, g : \mathbb{N} \to \mathbb{N}$ be two functions. We define $f \preceq g$, if there exist $i, j, k, l, m \in \mathbb{N}$ such that

$$f(n) \leq i g(jn + k) + ln + m$$

for all $n \in \mathbb{N}$. The functions $f$ and $g$ are called *equivalent* if $f \preceq g$ and $g \preceq f$. For example, if $f$ and $g$ are polynomials of the same positive degree, then they are equivalent. Without proof we cite here (see also Exercise 1):

**Lemma 10.10** *Isoperimetric functions of quasi-isometric, finitely presented groups are equivalent.*

Since the Cayley graphs of the same group are quasi-isometric according to Theorem 10.7, it follows that an isoperimetric function is independent of the presentation in which a group is given, up to equivalence. So it makes sense to say: The group $G$ has a linear (quadratic, exponential, etc.) isoperimetric inequality when its isoperimetric function is of the corresponding class. It is clear that finitely generated free groups have a linear isoperimetric function. A presentation of a free group (which has no relations) has the isoperimetric function $f(n) = 0, \forall n \in \mathbb{N}$, and this is linear.

A function is called *recursively computable* if there is a computer program that can calculate the image value for each pre-image. The following theorem shows that the word problem in a group is only undecidable if its isoperimetric function grows too fast.

**Theorem 10.11** *A finite presentation $P = \langle X \mid R \rangle$ has solvable word problem if and only if the associated isoperimetric function is recursively computable.*

***Proof*** Let the isoperimetric function $f$ of $P$ be recursively computable. Let $w$ be any word in $X$ of length $n$. If $w$ is a relation, there must therefore be a van Kampen diagram $\Gamma$ with at most $f(n)$ many disks. The number of these van Kampen diagrams is initially unlimited because there could be long 1-dimensional connecting pieces as in the van Kampen diagram of Fig. 10.3. However, the boundary of the van Kampen diagram has length $n$, so there are at most $n/2$ edges in $\Gamma$ that do not lie in the boundary of disks. But there are only finitely many of these van Kampen diagrams. We test all of them. Then $w$ is a relation if and only if one of them reads $w$ on its boundary.

We omit the proof of the converse.                                                    □

The algorithm for solving the word problem mentioned here is very ineffective if $f$ is a rapidly growing (e.g., exponentially growing) function. More effective algorithms for solving the word problem are usually found in such cases.

In the following, we present a class of presentations with solvable word problem. For this, we introduce the so-called *small cancellation conditions* $C(p)$, $T(q)$. The idea behind the abstract formulation of the conditions is simple:

The condition $C(p)$ will guarantee that each inner disk in a van Kampen diagram has at least valence $p$. The condition $T(q)$ ensures that each inner vertex has at least valence $q$. However, we want to read these conditions from the associated presentation. Formally:

Let $P = \langle x_1, \ldots, x_n \mid r_1, \ldots, r_m \rangle$ be a finite presentation. Each relation $r_i$ is *cyclically shortened*, i.e., $r_i$ is reduced and does not have the form $x_j^\epsilon w x_j^{-\epsilon}$. If $r_i$ were not cyclically shortened, $r_i$ could be made shorter with Tietze transformations.

The set of cyclically shortened relations $R = \{r_1, \ldots, r_m\}$ is called *symmetrized* if for each $r_i \in R$ also $r_i^{-1} \in R$ and in addition: If $r_i \in R$ is written arbitrarily as a product of two words $r_i = wv$, there must be a $r_j \in R$ with $r_j = vw$. If you have an arbitrary finite presentation, you can transform it into a finite, symmetrized set of relations by greatly enlarging the set of relators through Tietze transformations.

If $r_i = wv$ and $r_j = wu$ are different relations of $R$, then $w$ is called a *piece* relative to $R$. Given a van Kampen diagram without vertices of valence 1 over a symmetrized presentation, a connected part of the common boundary of two disks reads a piece if the two disks belong to different relations. For instance, we see from Fig. 10.5, that $j^{-1}k$ is a piece in the corresponding presentation.

**Definition 10.12** A finite presentation $P = \langle x_1, \ldots, x_n \mid r_1, \ldots, r_m \rangle$ fulfills the condition $C(p)$ if no relation of the associated symmetrized presentation can be written with fewer than $p$ pieces.

**Definition 10.13** A finite presentation $P = \langle x_1, \ldots, x_n \mid r_1, \ldots, r_m \rangle$ satisfies the condition $T(q)$ if for all $k$ with $3 \le k < q$ the following is fulfilled: Let $r_{i_1}, \ldots, r_{i_k}$ be relations of the corresponding symmetrized set of relations, such that $r_{i_j}^{-1} \neq r_{i_{j+1}}$. Then, in at least one of the products $r_{i_1}r_{i_2}, \ldots, r_{i_{k-1}}r_{i_k}, r_{i_k}r_{i_1}$ no reduction is possible.

Let $F$ be a connected finite graph in the Euclidean plane without crossing edges. Let $F$ divide the plane into $S$ regions (not counting the unbounded region), $K$ edges, and $V$ vertices. Each region is a disk (a hole in a region is forbidden). From $F$ one can calculate the *Euler characteristic* (see for example [Sti92]). It is given by the following formula:

$$\chi(F) = S - K + V.$$

Different parts of $F$, such as in the van Kampen diagram of Fig. 10.3, may be connected only through a path of edges and vertices. One easily proves:

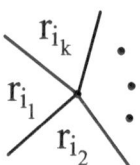

**Theorem 10.14** *If F is a connected finite graph in the Euclidean plane, then*
$\chi(F) = 1$.

**Theorem 10.15** *If* $P = \langle x_1, \ldots, x_n \mid r_1, \ldots, r_m \rangle$ *is a finite presentation that satisfies C(3),T(7) or C(4), T(5) or C(5),T(4) or C(7),T(3), then P satisfies a linear isoperimetric inequality.*

*Proof* Let $F$ be a van Kampen diagram without vertices of valence 2 over the presentation $P$ for a word $w$. If $P$ satisfies the condition $C(p)$, then every inner region of $F$ (i.e., every region without edges on the boundary of $F$) has at least valence $p$. Two relations have at most one piece in common. Each such piece can become an edge, and by Definition 10.12, each relation has at least $p$ pieces.

Similarly, if a presentation satisfies the condition $T(q)$, then every inner vertex of $F$ (i.e., a vertex not in the boundary of $F$) has valence $q$. If an inner vertex had valence $k < q$, then relations belonging to adjacent regions would have to shorten at common edges, and this is forbidden by Definition 10.13 (see Fig. 10.6).

Exemplarily, the case C(5),T(4) is proven here. The other cases are analogous.

$F$ has $S$ regions, $V$ vertices, and $K$ edges. Let $k$ be the number of edges on the boundary of $F$.

$2K - k$ counts all edges twice except those on the boundary, which are counted only once. Each region has valence at least 5, and therefore $5S$ gives the minimum number of edges on the boundary of regions twice, except those that lie on the boundary of $F$, which count only once. It follows that

$$5S \leq 2K - k. \tag{10.1}$$

There are at most $V - k$ inner vertices, and these all have, due to T(4), valence at least 4. This leads to

$$4(V - k) \leq 2(K - k) - \bar{K}, \tag{10.2}$$

where $\bar{K}$ denotes the number of edges with exactly one point on the boundary of $F$.
Equivalent to (10.2) is

$$4V - 2k \leq 2K - \bar{K}. \tag{10.3}$$

If we add the inequalities (10.1) and (10.3), we obtain

$$5S + 4V - 2k \leq 4K - k - \bar{K}$$

$$S + (4S - 4K + 4V) \leq k - \bar{K}.$$

From the Euler characteristic of $F$ (see Theorem 10.14) we now get

$$S + 4 \leq k - \bar{K}$$

and thus

$$S \leq k - \bar{K} - 4 \leq k - 4 \leq l(w) - 4.$$

Here, $l(w)$ is the length of $w$, and we obtain a linear isoperimetric inequality.    □

**Example 10.16** The presentation $\langle a, b, c, d \mid [a, b][c, d] \rangle$ is of type C(8), T(3), and thus has a linear isoperimetric function.

We check this with the help of GAP. We have to load the small cancellation package of I. SADOFSCHI COSTA (see [Cos18]). Then

```
gap> F:=FreeGroup(["a","b","c","d"]);;
gap> AssignGeneratorVariables(F);;
#I  Assigned the global variables [ a, b, c, d ]
gap> r:=a*b*a^-1*b^-1*c*d*c^-1*d^-1;;
gap> G:=F/[r];;
gap> PiecesOfGroup(G);
[ a^-1, a, b^-1, b, c^-1, c, d^-1, d ]
gap> GroupSatisfiesC(G,8);
true
gap> GroupSatisfiesT(G,3);
true
```

The pieces are exactly the generators and their inverses, and since the relator has length 8, C(8) is satisfied. We checked C(8) with the command

```
GroupSatisfiesC(G,8);
```

A decomposition of type $(n, m)$ essentially corresponds to the conditions $C(n)$ and $T(m)$: These are decompositions of the hyperbolic or Euclidean plane or the 2-sphere. Decompositions of type $(3, 7)$, $(4, 5)$, $(5, 4)$ and $(7, 3)$ are decompositions of the hyperbolic plane (see Theorem 9.7). This suggests that the presentations of Theorem 10.15 are in some sense "hyperbolic". We will clarify this fact in Sect. 10.4.

The following theorem is given without proof (a proof can be found, for example, in [HR93a]).

**Theorem 10.17** *If $P = \langle x_1, \ldots, x_n \mid r_1, \ldots, r_m \rangle$ is a finite presentation that satisfies C(3),T(6) or C(4), T(4) or C(6),T(3), then $P$ satisfies a quadratic isoperimetric inequality.*

The cases mentioned in this theorem correspond to the Euclidean decompositions. In fact, the relationship between boundary length and area of a region in the
Euclidean plane is quadratic. As an example, let $M_n$ be a square with edge length
$n$ in the Euclidean plane, consisting of $n$ times $n$ unit squares. This square satisfies
for all $n \in \mathbb{N}$ the condition C(4),T(4). Its boundary length $r$ is $4n$ and its area,
i.e., the number of its unit squares, is $n^2 = (r/4)^2$. So we have something like an
isoperimetric function: $f(r) = (r/4)^2$. More precisely:

**Example 10.18** The presentation $\langle a, b \mid ab = ba \rangle$ of the group $\mathbb{Z} \times \mathbb{Z}$ satisfies
the conditions $C(4)$, $T(4)$ and thus, according to Theorem 10.17, a quadratic
isoperimetric inequality.

The conditions C(4), T(4) can be clearly seen in the van Kampen diagram of
Fig. 10.1. The vertex in the middle has valence 4. Each generator is a piece in itself.

The relationship between area and boundary length in the hyperbolic plane is at
most linear. There are even triangles with infinite boundary length, but finite area
(see Fig. 9.5).

## Exercises

1. Prove directly (without the detour via quasi-isometry), that isoperimetric functions of two presentations of the same group are equivalent. (Hint: Observe the
   change of lengths under Tietze transformations.)
2. Prove Theorem 10.15 in the case C(4),T(5).
3. Prove that in the presentation of Example 10.16 pieces have length 1.
4. Check the presentation of the quaternion group of Example 10.4 with the help of
   GAP for pieces and C and T conditions.

## 10.4  Hyperbolic Groups

With the introduction of hyperbolic groups in 1987 by M. GROMOV (see [Gro87]),
there was a turning point in the development of geometric group theory. We want
to give a first introduction to the theory of hyperbolic groups here. The contents of
this section are mostly best represented in [Lö17], but also [A+91, Can02, dlH00,
BH99]. Gromov's original work [Gro87] is somewhat harder to read.

We recall the operation of a group on its Cayley graph of Theorem 4.52. If $g \in G$
is a group element and $x \in \Gamma_G$ is a vertex of the Cayley graph (which is nothing else
than a group element $x \in G$), then $g(x) = g \cdot x$, where $\cdot$ is the operation in the group
$G$. This is compatible with the operation on the edges, because if $k \in \Gamma_G$ is an edge
from the vertex $h'$ to $h$, which is labeled with $g_i$ (i.e., $h' \cdot g_i = h$), then $g(k) \in \Gamma_G$ is
an edge from the vertex $g \cdot h'$ to $g \cdot h$, which is labeled with $g_i$. With the word metric,
the operation of a group element $g$ on the Cayley graph is distance-preserving and

thus an isometry. Every finitely generated group is therefore a symmetry group of a metric space, namely its Cayley graph.

In Sect. 9.2 we defined for the hyperbolic plane what it means to be thin. We generalize this notion to arbitrary geodesic metric spaces $(X, d)$: $(X, d)$ is called $\delta$-*hyperbolic* if there is a constant $\delta \in \mathbb{R}$ such that the following holds: Let $a, b, c \subset X$ be sides of a triangle. Then every point $P$ on $a$ has distance at most $\delta$ to $b$ or to $c$. The triangle is then called $\delta$-*thin* (see Fig. 9.6 on page 169).

**Definition 10.19** Let $\delta \in \mathbb{R}$ be a constant. The finitely generated group $G$ is called $\delta$-*hyperbolic*, *wordhyperbolic* or *hyperbolic*, if it has a finite generating system $\{g_1, \ldots, g_n\}$ such that every triangle in the associated Cayley graph $\Gamma_G(g_1, \ldots, g_n)$ is $\delta$-thin.

A proof of the following theorem can be found in [Lö17].

**Theorem 10.20** *Let* $(X, d_X)$ *and* $(Y, d_Y)$ *be quasi-isometric, geodesic, metric spaces. If* $(X, d_X)$ *is* $\delta$-*hyperbolic, then there is a constant* $\delta'$ *such that* $(Y, d_Y)$ *is* $\delta'$-*hyperbolic.*

Since Cayley graphs of the same group are quasi-isometric (see Theorem 10.7), it follows that the definition of a hyperbolic group is independent of the chosen finite generating system, i.e., if every triangle is $\delta$-thin in the Cayley graph of a group with respect to a finite generating system $\{g_1, \ldots, g_n\}$, then for any other finite generating system $\{h_1, \ldots, h_m\}$ of the same group there is a constant $\delta'$ such that every triangle is $\delta'$-thin in the Cayley graph corresponding to $\{h_1, \ldots, h_m\}$. Being hyperbolic is therefore a property of a group.

Every finite group is hyperbolic. As the constant $\delta$, one can simply take the maximum over all distances between any two vertices in the Cayley graph.

According to Theorem 5.10, the Cayley graph of a free group with a suitable generating system is a tree. However, every tree is 0-thin: a triangle shrinks to a tripod, as shown in Fig. 10.7.

So we have proven:

**Theorem 10.21** *Finitely generated free groups are hyperbolic.*

$\mathbb{Z} \times \mathbb{Z}$ is not hyperbolic: The subgroup of translations of a decomposition of the plane into unit squares is $\mathbb{Z} \times \mathbb{Z}$ (compare Sect. 6.4). Let $a, b$ be generators of this group, corresponding to translations by length 1 in the vertical and horizontal directions. The associated Cayley graph is the lattice consisting of all horizontal and

**Fig. 10.7** Triangle in a tree

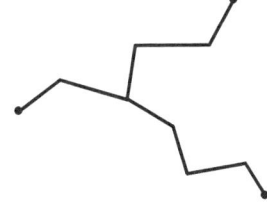

**Fig. 10.8** Triangle in the
Cayley graph of $\mathbb{Z} \times \mathbb{Z}$

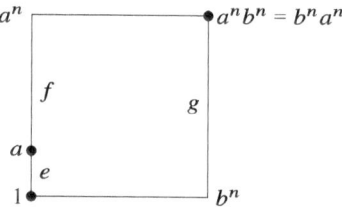

vertical lines in the plane through integer coordinate points. For each number $n \in \mathbb{N}$, we consider in this Cayley graph the triangle consisting of the sides

- $e$ given by the word $a$,
- $f$ from $a$ to $a^n b^n$ given by the word $a^{n-1} b^n$ and
- $g$ given by the word $b^n a^n$

(see Fig. 10.8). Even though it doesn't look like it in Fig. 10.8, these are indeed triangles. The edges are all three geodesics. As $n$ increases, the distance from the vertex $b^n$ to the other two edges $e$, $f$ becomes arbitrarily large, and therefore there is no constant $\delta$ for all these triangles.

The fact that $\mathbb{Z} \times \mathbb{Z}$ is not hyperbolic is not surprising. The group comes from a decomposition of the Euclidean plane. The Cayley graph is naturally embedded in the Euclidean plane, and there is no constant $\delta$ such that all triangles are $\delta$-thin, as justified in Sect. 9.2. It can even be proven that if a group contains $\mathbb{Z} \times \mathbb{Z}$ as a subgroup, it is not hyperbolic.

**Theorem 10.22** *Assume $H < G$ is a subgroup of finite index of the group $G$ and $H$ and $G$ are finitely generated. Then $G$ is hyperbolic if and only if $H$ is hyperbolic.*

**Proof** Observe that the inclusion of the vertices of a Cayley graph of $H$ into a Cayley graph of $G$ is a quasi-isometry. Then Theorem 10.20 gives the desired result. $\square$

A group is called *virtually cyclic* if it contains $\mathbb{Z}$ as a subgroup of finite index. Since $\mathbb{Z}$ is hyperbolic by Theorem 10.21, we know from Theorem 10.22 that virtually cyclic groups are hyperbolic. The infinite dihedral group (see Example 6.11) is virtually cyclic and hence hyperbolic. In order to see that $D_\infty = \langle x, y \mid x^2, y^2 \rangle$ has $\mathbb{Z}$ as a subgroup of finite index we construct a homomorphism $\phi \colon D_\infty \to \mathbb{Z}_2 \times \mathbb{Z}_2$ by mapping $x \to (1, 0)$ and $y \to (0, 1)$. The kernel of $\phi$ is generated by the commutator $[x, y]$ and is hence isomorphic to $\mathbb{Z}$. The image is a group of order 4 and hence finite. So there are 4 cosets and the index of $\mathbb{Z}$ in $D_\infty$ is finite.

Further examples of hyperbolic groups are those of the decompositions of Theorem 9.7 on page 173. Each of these groups and the underlying triangle groups come from a decomposition of the hyperbolic plane, so each Cayley graph is naturally embedded in the hyperbolic plane, and we can conclude:

**Theorem 10.23** *If $\frac{1}{m} + \frac{1}{n} < \frac{1}{2}$, then the group $G_{(n,m)}$ of the decomposition of the hyperbolic plane of type $(n, m)$ is hyperbolic.*

**Proof** Let $\Gamma$ be the Cayley graph of the group $G_{(n,m)}$ for natural numbers $n, m$ for which $\frac{1}{m} + \frac{1}{n} < \frac{1}{2}$. According to the ŠVARC–MILNOR theorem, Theorem 10.8, $\Gamma$ is quasi-isometric to the hyperbolic plane. This is $\delta$-hyperbolic according to Theorem 9.3, and due to Theorem 10.20, $\Gamma$ is also $\delta$-hyperbolic. $\qquad\square$

There is a very effective algorithm for solving the word problem in hyperbolic groups, the so-called Dehn algorithm. It is named after MAX DEHN, who used it to solve the word problem for surface groups in several papers at the beginning of the last century.

If $w = uv$ is a relation in a group consisting of the subwords $u, v$, then $w' = u^{-1} \cdot uv \cdot u = vu$ is also a relation. We say $w'$ arises from $w$ through *cyclic conjugation*.

**Definition 10.24** A *Dehn presentation* is a finite presentation in which every nontrivial reduced relation (after possible inversion and cyclic conjugation) contains a subword that is found in a defining relation $r$ and is longer than half of $r$.

A finite presentation $\langle X \mid R \rangle$ is thus a Dehn presentation if for all reduced relations $w$ in the generators and their inverses with $l(w) > 0$ there is a defining relation $r = r_1 r_2 \in R'$ of the symmetrized relator set $R'$ of $R$ with $w = u r_1 v$ and $l(r_1) > l(r_2)$.

If a group has a Dehn presentation $\langle X \mid R \rangle$, we can solve the word problem with the following *Dehn algorithm*: We consider all subwords of a given word $w$. If a subword $r_1$ is found in $w = u r_1 v$ which occurs in a relation $r = r_1 r_2$ and is more than half as long as the relation, we replace $w = u r_1 v$ with $w' = u r_2^{-1} v$ and have found a shorter word that corresponds to the same group element. If such a subword cannot be found in $w$, then $w$ is not a relation, and we stop. But if we do find such a subword, we continue with $w'$ after possible free reductions inductively. If we end with the empty word, then $w$ is a relation. If we stop after any partial step, then $w$ is non-trivial in the group.

With this algorithm, the word problem is very efficiently solvable. So efficient that you generally don't need a computer for it.

**Example 10.25** The presentation $\langle a, b, c, d \mid [a, b][c, d] \rangle$ of Example 10.16 is a Dehn presentation.

Dehn's original proof shows that the Cayley graph of this group lies as a decomposition of type (8,8) in the hyperbolic plane. Given a relation $r$ in this group there is a path $w$ in the Cayley graph that belongs to $r$. $w$ has a maximum distance from the vertex 1. This is where the subword of $w$ is, which contains more than half of the defining relation (see also Exercise 3).

**Theorem 10.26** *If a group has a Dehn presentation, then it has a linear isoperimetric function.*

**Proof** When shortening a subword of a relation $w$ by the shorter part of a defining relation, the number of relations in a van Kampen diagram for $w$ decreases by one and $w$ becomes at least one shorter. This results in the isoperimetric function $f(n) = n$. $\qquad\square$

**Theorem 10.27**  *Let G be a δ-hyperbolic group with finite generating system X. Let R be the set of all relations in G of length at most 16δ + 1. Then ⟨X | R⟩ is a Dehn presentation for G.*

It can be shown that even the set of all relations of length at most 8δ is sufficient for a Dehn presentation (see [A⁺91]).

If $\delta = 0$, then $G$ is a free group with free generating system $X$. In this case, the statement of the theorem follows. We therefore assume in the following that $\delta \geq 1$. The proof of Theorem 10.27 works with length estimates in the Cayley graph. We again identify vertices of the Cayley graph with group elements and words in the generators (and their inverses) with paths in the Cayley graph from 1 to the corresponding group element.

**Definition 10.28**  A path $w$ in the Cayley graph is a *k-local geodesic* if every subpath of length at most $k$ is a geodesic.

With $d(P, Q)$ we denote the distance from a point $P$ to a point $Q$ in the Cayley graph with respect to the word metric. If $w$ is a path and $P$ a point, then $d(P, w)$ is the shortest distance from $P$ to any point on $w$.

To prove Theorem 10.27 we need the following lemma:

**Lemma 10.29**  *Let G be a δ-hyperbolic group and u a k-local geodesic, where k > 8δ. Let v be a geodesic that ends at the same group element g as u. Let P be any point on u. Then d(P, v) ≤ 2δ.*

**Proof**  Let $P \in u$ be a point of greatest distance to $v$. We first assume that $d(1, P) > 4\delta$ and $d(P, g) > 4\delta$. Let $Q$ be a point on $u$ between 1 and $P$, such that $k/2 > d(Q, P) > 4\delta$. Similarly, let $S$ be a point on $u$ between $P$ and $g$, such that $k/2 > d(S, P) > 4\delta$ (see Fig. 10.9).

Let $Q'$ be the point on $v$ with the smallest distance to $Q$ and $S'$ the point on $v$ with the smallest distance to $S$. We consider the square $Q, S, S', Q'$. We draw the diagonal $QS'$ into this square and obtain 2 triangles. Now we apply the criterion for δ-thin triangles to the upper triangle from $P$. First we consider the case where $P$ has distance at most $\delta$ to a point $P'$ on $QS'$. The case where this point lies on $SS'$ is treated later.

Similarly, we apply to the triangle $QQ'S'$ from $P'$ the criterion for δ-thin triangles and thus obtain another point $T$ on $QQ'$ or on $Q'S'$ with distance at most $\delta$ from $P'$. If we can show that $T$ lies on $Q'S'$, then $P$ will have at most distance $2\delta$

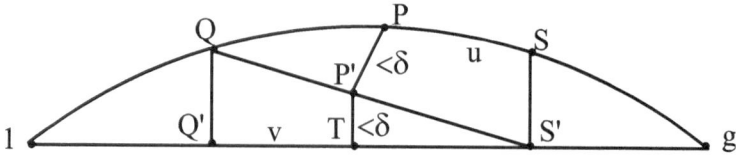

**Fig. 10.9**  A geodesic and a $k$-local geodesic

to $T$ and thus to $v$, and the statement will be proven. If $T$ lies on $QQ'$, we obtain a contradiction in the following way:

$$d(P, Q') - d(Q, Q') \le \big(d(P, T) + d(T, Q')\big) - \big(d(Q, T) + d(T, Q')\big)$$
$$= d(P, T) - d(Q, T)$$
$$\le d(P, T) - (d(Q, P) - d(P, T))$$
$$= 2d(P, T) - d(P, Q)$$
$$< 4\delta - 4\delta = 0.$$

It follows that $d(P, Q') < d(Q, Q') = d(Q, v)$, which contradicts the assumption that $P$ is a point of greatest distance to $v$. So $T$ lies on $Q'S'$, and $P$ has a distance of at most $2\delta$ to $T$ and thus to $v$.

In the case when $P'$ lies on $SS'$, an analogous calculation (in the above, replace $T, Q, Q'$ by $P', S, S'$) gives a contradiction. If $d(1, P) \le 4\delta$ or $d(P, g) \le 4\delta$, similar arguments lead to the proof. In this case, $P$ even has only the distance $\delta$ to $v$.                                                                                                         □

***Proof of Theorem 10.27*** Let $w$ be a relation and $k = 8\delta + 1$. If $w$ is not a $k$-local geodesic, then $w$ has a subpath of length at most $8\delta + 1$, which is not a geodesic. The corresponding geodesic has a length of at most $8\delta$ and, together with the subpath, forms a relation that, due to its length restriction, is in the given presentation. Thus, $w$ can be shortened.

Now let $w$ be a $k$-local geodesic. Since $w$ is a relation, the path is closed, and the corresponding geodesic is the constant path that stays on 1. According to Lemma 10.29, every point on $w$ has a distance of at most $2\delta$ from 1. But then $w$ has at most length $2\delta$, because if $w$ had a subpath, starting at 1, of length $2\delta + 1$, then this subpath would be a geodesic and would have at its end a vertex at distance $2\delta + 1$ from 1. Contradiction!                                                                       □

From Theorem 10.27 we immediately get the following:

**Corollary 10.30** *Hyperbolic groups are finitely presented.*

Overall, we have the following theorem (see [A$^+$91, BH99]):

**Theorem 10.31** *For a finitely generated group $G$, the following statements are equivalent:*

1. *$G$ is hyperbolic.*
2. *$G$ satisfies a linear isoperimetric inequality.*
3. *$G$ has a Dehn presentation.*

It then follows from Theorem 10.15 that groups with finite presentations that satisfy one of the small cancellation conditions C(3),T(7) or C(4), T(5) or C(5),T(4) or C(7),T(3) are hyperbolic.

## Exercises

1. Are the frieze groups hyperbolic?
2. Prove that the operation of an element of a finitely generated group on its Cayley graph is an isometry.
3. Prove the statement of Example 10.25.

## 10.5  Combings

In certain cases, quadratic isoperimetric inequalities can be specified for presentations. One of the simplest ways to do this is to specify a normal form of words for the group with certain properties, which is called combing (see also [E$^+$92]).

Before we do that, we need a more detailed description of paths in the Cayley graph. If $G$ is a group with a finite generating system $X = \{x_1, \ldots x_n\}$ and $g \in G$, we consider a normal form $w_g$ in the generators and their inverses. Each normal form corresponds to a path—which we also want to call $w_g$—in the Cayley graph from the vertex 1 to the vertex $g$. More precisely: $w_g$ is a continuous mapping $w_g \colon \mathbb{R}^+ \to \Gamma_G$. We want the path to end in $g$ and stay there for large $t$: So there is an $r \in \mathbb{R}^+$ such that $w_g(t) = g$ for all $t \geq r$.

**Definition 10.32** Let $G$ be a group with a finite generating system $X$. If there are constants $c, k \in \mathbb{N}$ and for each element $g \in G$ a path $w_g \in \Gamma_G$ in the Cayley graph between 1 and $g$ satisfying

1. $|w_g| \leq c \cdot d(1, g)$, and
2. if $w_g(t) \in \Gamma_G$ is a vertex on the path $w_g$, then for each generator and its inverse $x \in X^{\pm 1}$ the inequality $d(w_g(t), w_{gx}(t)) \leq k$ holds,

then $G$ is said to be *combed* with respect to $X$ (see Fig. 10.10).

As an example, we again consider the subgroup of translations of the decomposition of the plane into unit squares with the generators $a, b$, which correspond to the translations of length 1 in horizontal and vertical direction. This is the group $\mathbb{Z} \times \mathbb{Z}$ presented by $\langle a, b \mid ab = ba \rangle$.

**Fig. 10.10**  Distance
between paths in the Cayley
graph

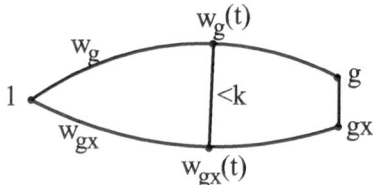

If we choose the normal form of the words so that we first perform the translations in the $a$- and then in the $b$-direction, we get geodesics (so we can choose $c = 1$), and these are close to each other (choose $k = 1$).

If $G$ is combable with respect to the finite generating system $X$, then $G$ is combable with respect to any other finite generating system. Indeed, the following holds:

**Theorem 10.33** *If the Cayley graphs $\Gamma_G(x_1, \ldots x_n)$ and $\Gamma_H(y_1, \ldots, y_m)$ of the groups $G$ and $H$ are quasi-isometric and if $G$ is combable with respect to $\{x_1, \ldots x_n\}$, then $H$ is combable with respect to $\{y_1, \ldots, y_m\}$.*

**Proof** If a quasi-isometry $f : \Gamma_G(x_1, \ldots x_n) \to \Gamma_H(y_1, \ldots, y_m)$ is given, then the constants in $\Gamma_G$ can easily be expressed in terms of those in $\Gamma_H$ and vice versa.  □

So we can talk about whether groups are combable or not. We have just seen that the following holds:

**Theorem 10.34** *The group $\mathbb{Z} \times \mathbb{Z}$ is combable.*

Combings have been introduced because of the following theorem:

**Theorem 10.35** *If the group $G$ is combable, then it is finitely presented and has a quadratic isoperimetric inequality.*

**Proof** If $v$ is a word in the finitely many generators $X = \{x_1, \ldots x_n\}$ and their inverses, then this word corresponds to a path $v \in \Gamma_G$ starting at the vertex 1. If $v = 1$ in $G$, then the path also ends at the vertex 1. If $v$ has length $m$ in the word metric, then we call the vertices that $v$ traverses $1 = P_1, \ldots, P_m = 1$. We consider the paths $w_{P_i}$ and connect adjacent vertices $w_{P_i}(t)$ and $w_{P_{i+1}}(t)$ by geodesics in the Cayley graph (see Fig. 10.11). Because $G$ is combable, these connections have length at most $k$. This results in closed paths (in the picture the small rectangles) of length at most $2k + 2$. If $R$ is therefore the set of all words of length at most $2k + 2$ that are trivial in $G$, then $P = \langle X \mid R \rangle$ is a finite presentation for $G$. Each relation can be trivialized like $v$ through these short relations.

Figure 10.11 gives us a van Kampen diagram for $v$. We estimate the number of its disks from above. $m$ is the length of $v$. Let $v_i$ be the word consisting of the first $i$ letters of $v$. $c$ is the constant from condition 1. of Definition 10.32. Then for all $i$:

**Fig. 10.11**  Combing of a word $v = 1$ in $G$

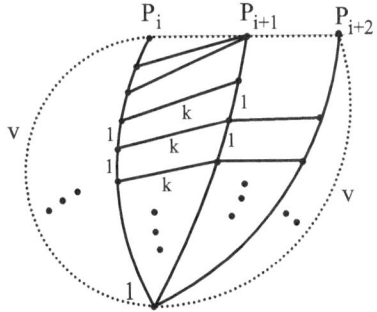

$|w_{P_i}| \leq c \cdot d(1, v_i) \leq c \cdot m/2$. There are $m$ words $v_1, \ldots, v_m$, and therefore in the diagram of Fig. 10.11 there are at most $m \cdot cm/2 = c \cdot m^2/2$ many disks. This gives us a quadratic isoperimetric function for $G$.                                                $\square$

A class of combable groups is very easy to find:

**Theorem 10.36** *Hyperbolic groups are combable.*

**Proof** If one takes geodesics in the Cayley graph as normal forms for group elements, then two group elements $g$ and $gx$ with distance 1 together with the vertex 1 form a triangle. Because the group is $\delta$-hyperbolic, the paths $w_g$ and $w_{gx}$ are at every point at most $\delta + 1$ apart from each other. The bound $\delta + 1$ can only be assumed at the point $P$ of the path $w_g$ that is $\delta$ away from $g$. All points of $w_g$ that are closer to 1 than $P$ have distance at most $\delta$ from $w_{gx}$. All points $Q$ of $w_g$ that are closer to $g$ than $P$ have, in the path from $Q$ through $g$ to $gx$, distance at most $\delta$ from $w_{gx}$.   $\square$

In [E$^+$92] and [HR93b] one can find weakened combing concepts, which lead to isoperimetric inequalities that are no longer quadratic.

## Exercises

1. Prove Theorem 10.33 in detail by carefully calculating the constants that arise.
2. Prove that $\mathbb{Z} \times \mathbb{Z} \times \mathbb{Z}$ is combable.

# Appendix A
# The Isometries of the Plane

Here we catch up on the proof of a theorem of the first chapter:

**Theorem 1.6** An isometry of the plane with a fixed point is a rotation if it preserves the orientation. It is a reflection if it does not preserve the orientation. An isometry of the plane without a fixed point is a translation if it preserves the orientation, and otherwise a glide reflection.

***Proof*** Let $f$ be an isometry with fixed point $P$. Since $f$ preserves distances, it maps every circle with center $P$ onto itself. A circle only allows rotations and reflections as isometries onto itself (this should be clear at this point), and thus $f$ is a rotation or a reflection.

From now on, let $f$ be without fixed points. Then $f^2$ also has no fixed point because if $f^2(P) = P$ for some point $P$, then the isometry $f$ would swap the point $P$ with $f(P)$, and the midpoint $A$ of the interval $[P, f(P)]$ would be a fixed point of $f$ since we have

$$\overline{f(P), f(A)} = \overline{P, A} = \overline{f(P), A} = \overline{P, f(A)}.$$

Let $f$ additionally preserve orientation and $P$ be any point. If the lines $\overline{P, f(P)}$ and $\overline{f(P), f^2(P)}$ are not parallel, then the two perpendiculars to the midpoints of these lines intersect at a point $Q$ (see Fig. A.1). $f$ maps the entire line $\overline{P, f(P)}$ onto $\overline{f(P), f^2(P)}$ and, because $f$ preserves orientation, $Q$ onto itself, and this contradicts the absence of fixed points. Therefore, the lines $\overline{P, f(P)}$ and $\overline{f(P), f^2(P)}$ are parallel. Thus, the points $P$, $f(P)$, $f^2(P)$ all lie on a straight line $g$ and are all different. Then all $f^k(P)$ lie on $g$. $f$ thus operates on this line by translation.

We choose a point $P' \notin g$ and apply the same arguments to $P'$ as we have just done to $P$. We find a line $g'$ such that all $f^k(P')$ lie on $g'$. If $g$ and $g'$ had an intersection point $S$, then $S$ would have a preimage on $g$ and one on $g'$ under $f$ and

S. Rosebrock, *Visual Group Theory*, Springer Undergraduate Mathematics Series, https://doi.org/10.1007/978-3-662-69365-0

**Fig. A.1** $f$ preserving
orientation and without fixed
points

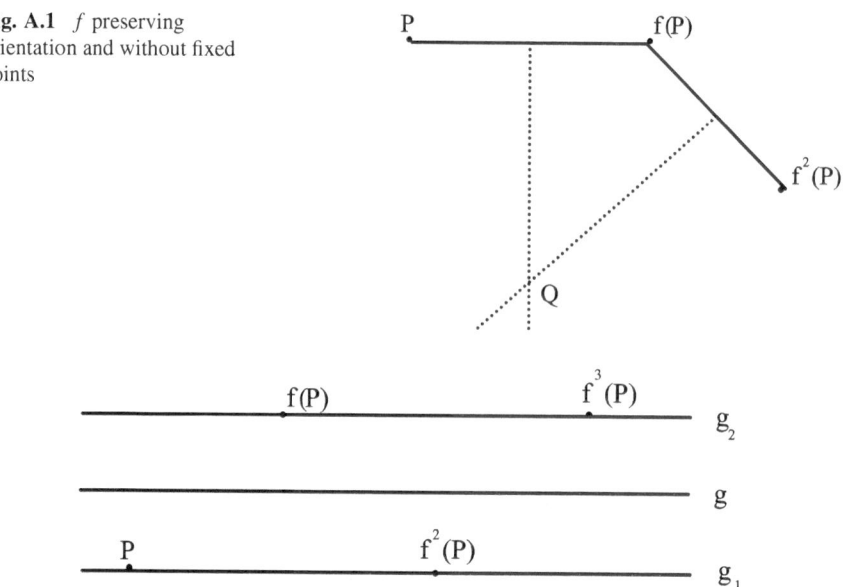

**Fig. A.2** $f$ reversing orientation and without fixed points

would not be injective. Therefore, $g'$ is parallel to $g$, and $f$ acts on $g'$ by the same translation. Thus, it acts on the entire plane by translation.

If $f$ is without fixed points and reverses orientation, then $f^2$ preserves orientation. The arguments just given, applied to $f^2$, show that $f^2$ is a nontrivial translation. Let $P$ be any point. $f^2$ leaves the line $g_1$, through the points $P$ and $f^2(P)$, as well as the line $g_2$, through the points $f(P)$ and $f^3(P)$, invariant. Therefore, $g_1$ and $g_2$ are parallel (and not necessarily different). $f$ maps $g_1$ onto $g_2$ and vice versa (see Fig. A.2).

Thus, the line $g$ between $g_1$ and $g_2$ is invariant under $f$. Since $f^2$ acts as a translation on $g$, $f$ also operates as translation $t$ on $g$. Since $f$ is orientation-reversing, it follows that $f = t \circ s_g$. Thus, $f$ is a glide reflection.                □

# Appendix B
# Matrices

Understanding this section requires knowledge of linear algebra. Matrix groups are important examples of groups, and many authors use them. Many of the groups discussed in the book can also be described as matrix groups, so we want to give a brief introduction to matrix groups.

Given two linear mappings $g, f : \mathbb{R}^n \to \mathbb{R}^n$, you can execute them one after the other and, as for mappings in general, this composition is associative. If you only consider all invertible linear mappings of $\mathbb{R}^n$ onto itself, these form a group, the *linear group*, abbreviated $GL(n, \mathbb{R})$. Conveniently, linear mappings are represented by matrices. If $f : \mathbb{R}^n \to \mathbb{R}^n$ is an invertible linear mapping, there is an invertible $n \times n$ matrix $A$ with real entries which describes $f$ with respect to the standard basis, consisting of unit vectors. If $\vec{x}$ is thus a vector in $\mathbb{R}^n$, then $f(\vec{x})$ is described by $A\vec{x}$.

The group $GL(n, \mathbb{R})$ can therefore also be described as the group of invertible $n \times n$ matrices with real entries with respect to matrix multiplication. The identity element is the $n \times n$ identity matrix:

$$I_n = \begin{pmatrix} 1 & 0 & \cdots & 0 \\ 0 & 1 & \cdots & 0 \\ \vdots & \vdots & & \vdots \\ 0 & 0 & \cdots & 1 \end{pmatrix}.$$

The composition of linear mappings corresponds exactly to the multiplication of matrices, as is common in linear algebra. For $n = 1$, it is the multiplication of invertible real numbers. The group $GL(1, \mathbb{R})$ is thus isomorphic to the group $\mathbb{R} - \{0\}$ with the usual multiplication. For $n \geq 2$, the groups $GL(n, \mathbb{R})$ are not commutative.

Isometries of $\mathbb{R}^n$ onto itself, which fix the origin, are invertible linear mappings. These include reflections through hyperplanes passing through the origin and rotations about subspaces of dimension $n - 2$ through the origin. In the Euclidean

© The Author(s), under exclusive license to Springer-Verlag GmbH, DE, part of Springer Nature 2024
S. Rosebrock, *Visual Group Theory*, Springer Undergraduate Mathematics Series, https://doi.org/10.1007/978-3-662-69365-0

plane, these are reflections over lines through the origin and rotations about the origin.

For example, the matrix

$$\begin{pmatrix} 0 & 1 \\ 1 & 0 \end{pmatrix}$$

describes the reflection over the bisector in the Euclidean plane because

$$\begin{pmatrix} 0 & 1 \\ 1 & 0 \end{pmatrix} \begin{pmatrix} 1 \\ 0 \end{pmatrix} = \begin{pmatrix} 0 \\ 1 \end{pmatrix} \text{ and } \begin{pmatrix} 0 & 1 \\ 1 & 0 \end{pmatrix} \begin{pmatrix} 0 \\ 1 \end{pmatrix} = \begin{pmatrix} 1 \\ 0 \end{pmatrix}.$$

An invertible $n \times n$ matrix $A$ is called *orthogonal* if its transpose $A^t$ is equal to its inverse $A^{-1}$. So if you multiply the $i$-th row $\vec{a}$ of $A^t$ by the $i$-th column of $A$, this must yield $\vec{a} \cdot \vec{a} = 1$. The column vectors of $A$ are therefore vectors of length 1. If you multiply the $j$-th row of $A^t$ by the $i$-th column of $A$ for $i \neq j$, this must yield 0. The column vectors of $A$ are therefore orthogonal to each other. Since $\det A^t = \det A$ and $\det A^t \cdot \det A = 1$, the determinant $\det A$ of an orthogonal matrix $A$ is always $\pm 1$.

The subgroup of the linear group $GL(n, \mathbb{R})$ which consists only of orthogonal matrices is the *orthogonal group $O_n$*, which is already known to us. It is the stabilizer of the origin in the symmetry group of $\mathbb{R}^n$. It is indeed a subgroup, because if $A$ and $B$ are orthogonal, then $AB^{-1}$ is also orthogonal, as can be easily calculated, and according to Theorem 3.9 on page 35 it follows that $O_n < GL(n, \mathbb{R})$. The subgroup of $O_n$ of orientation-preserving isometries, i.e., the matrices with determinant $+1$, is the *special orthogonal group $SO_n$*. So far, we have denoted it as $O_n^+$.

In the Euclidean plane, only the rotations about the origin are orientation-preserving isometries that fix the origin. The corresponding matrices have the form

$$\begin{pmatrix} \cos \phi & -\sin \phi \\ \sin \phi & \cos \phi \end{pmatrix}.$$

This follows because rotations are orthogonal mappings (the column vectors are therefore on the unit circle and are orthogonal to each other). Therefore

$$SO_2 = \left\{ \begin{pmatrix} \cos \phi & -\sin \phi \\ \sin \phi & \cos \phi \end{pmatrix} \mid 0 \leq \phi < 2\pi \right\}.$$

Many of the groups discussed can be found again as matrix groups. For example, the group $D_3$ of the equilateral triangle with vertices $(1, -\sqrt{3})$, $(1, \sqrt{3})$, $(-2, 0)$ is generated by the matrices

$$\begin{pmatrix} 1 & 0 \\ 0 & -1 \end{pmatrix} \text{ and } \begin{pmatrix} -1/2 & \frac{\sqrt{3}}{2} \\ \frac{\sqrt{3}}{2} & 1/2 \end{pmatrix},$$

which correspond to a reflection over the $x$-axis and over the line through the points $(1, \sqrt{3})$ and $(0, 0)$.

In GAP we consider the group $D_4$ generated by the reflection a over the $x$-axis and the reflection b over the bisector. Matrices are notated in GAP as [*row 1, row 2,...*], where each row is a list of comma-separated values enclosed in square brackets.

```
gap> a:= [ [ 1, 0 ], [ 0, -1] ];;
gap> b:= [ [ 0, 1 ], [ 1,  0] ];;
gap> D4:=Group(a,b);
Group([ [ [ 1, 0 ], [ 0, -1 ] ], [ [ 0, 1 ], [ 1, 0 ] ] ])
gap> Order(D4);
8
gap> Elements(D4);
[ [ [ -1, 0 ], [ 0, -1 ] ],   [ [ -1, 0 ], [ 0, 1 ] ],
  [ [ 0, -1 ], [ -1, 0 ] ],   [ [ 0, -1 ], [ 1, 0 ] ],
  [ [ 0, 1 ], [ -1, 0 ] ],    [ [ 0, 1 ], [ 1, 0 ] ],
  [ [ 1, 0 ], [ 0, -1 ] ],    [ [ 1, 0 ], [ 0, 1 ] ] ]
```

Sometimes, algebraic and even geometric facts can be more easily understood using matrices than through other methods. We demonstrate this with an example:

**Theorem** If $G$ is a subgroup of the group $O_n$ which contains a point reflection $g$, then $g$ lies in the center of $G$.

**Proof** The point reflection must occur at the origin. Each coordinate is mapped to its negative by $g$. The point reflection can therefore be represented as a matrix in the following form:

$$-I_n = \begin{pmatrix} -1 & 0 & \cdots & 0 \\ 0 & -1 & \cdots & 0 \\ \vdots & \vdots & & \vdots \\ 0 & 0 & \cdots & -1 \end{pmatrix}.$$

It is easy to see that $-I_n$ commutes with every other matrix of $O_n$ because for $A \in O_n$ we have $(-I_n) \cdot A = -A = A \cdot (-I_n)$.                                             □

We give the proof of Theorem 4.19.

**Theorem 4.19** An orientation-preserving isometry in $\mathbb{R}^3$ which fixes the origin is a rotation about a line through the origin.

**Proof** Let $A$ be the matrix of an orientation-preserving isometry in $\mathbb{R}^3$ which fixes the origin, i.e., $A \in SO_3$. The characteristic polynomial, i.e., the determinant $\det(A - \lambda I_3)$, is a cubic polynomial, so it has (at least) one real root. $A$ therefore has a real eigenvalue. This eigenvalue must be $+1$ since the product of the eigenvalues must give the determinant of $A$ and $A \in SO_3$. The linear mapping associated with $A$ can therefore be represented by a matrix of the form

$$\begin{pmatrix} 1 & 0 & 0 \\ 0 & a & b \\ 0 & c & d \end{pmatrix}$$

after a change of basis. Here, $\begin{pmatrix} a & b \\ c & d \end{pmatrix} \in SO_2$, and this matrix can be interpreted as a rotation in a plane through the origin in $\mathbb{R}^3$. The axis of rotation runs perpendicular to this plane at the origin after the change of basis.                                                   □

An $n \times n$ matrix that only has entries 0 and 1 and exactly one 1 in each column and each row permutes a vector of length $n$. An example in the case $n = 3$:

$$\begin{pmatrix} 1 & 0 & 0 \\ 0 & 0 & 1 \\ 0 & 1 & 0 \end{pmatrix} \cdot \begin{pmatrix} a \\ b \\ c \end{pmatrix} = \begin{pmatrix} a \\ c \\ b \end{pmatrix}. \tag{B.1}$$

The set of all these $n \times n$ matrices thus corresponds to the set of all permutations of an $n$-tuple. Two such matrices combine according to the combination of permutations, so that we overall have a representation of the symmetric group $S_n$ as a matrix group:

**Theorem** The following holds:

$S_n \cong \{A \in GL(n, \mathbb{R}) \mid a_{ij} \in \{0, 1\},$ exactly one 1 in each column and each row$\}$.

According to Theorem 4.2, the group $S_n$ can be generated by transpositions. The matrix corresponding to the transposition $(i, j)$ is obtained by swapping the $i$-th row with the $j$-th row in the identity matrix. The matrix of (B.1) corresponds to the transposition $(2,3)$ in the group $S_3$.

According to Theorem 4.3, every finite group is a subgroup of one of the groups $S_n$. Therefore:

**Theorem** Every finite group $G$ with $|G| = n$ is a subgroup of the group $GL(n, \mathbb{R})$.

**Theorem** The group $GL(n, \mathbb{Z})$, i.e., the invertible $n \times n$ matrices with entries in $\mathbb{Z}$, is residually finite.

**Proof** Let $A \in GL(n, \mathbb{Z})$. Choose a prime number $p$ that is larger than any entry of $A$. Then $A$ is nontrivial in $GL(n, \mathbb{Z}_p)$. $GL(n, \mathbb{Z}_p)$ is a finite quotient of $GL(n, \mathbb{Z})$.                                                   □

# Appendix C
# List of Symbols

| Symbol | Explanation |
| --- | --- |
| $\mathbb{N}$ | The natural numbers |
| $\mathbb{Z}$ | The integers |
| $\mathbb{Z}_n$ | The numbers $\{0, \ldots, n-1\}$ |
| $n\mathbb{Z}$ | The integers divisible by $n$ |
| $\mathbb{Q}$ | The rational numbers |
| $\mathbb{R}^2$ | The Euclidean plane |
| $\mathbb{R}^n$ | The Euclidean space of dimension $n$ |
| $\mathbb{H}^2$ | The hyperbolic plane |
| $\mathbb{R}$ | The real numbers |
| $\in$ | Element of |
| $\notin$ | Not an element of |
| $\mid$ | Divides |
| $\subset$ | Subset of |
| $\subsetneq$ | Proper subset of |
| $\cup$ | Union |
| $\cap$ | Intersection |
| $\Rightarrow$ | Implies |
| $\Leftrightarrow$ | If and only if |
| $\forall$ | For all |
| $\exists$ | There exists |
| $\prod w_i$ | Product of the $w_i$ |
| $!$ | Factorial |
| $\times$ | Cartesian product (see Definition 6.1) |

(continued)

S. Rosebrock, *Visual Group Theory*, Springer Undergraduate Mathematics Series, https://doi.org/10.1007/978-3-662-69365-0

| Symbol | Explanation |
|--------|-------------|
| $\wedge$ | And |
| $\vee$ | Or |
| $\binom{n}{k}$ | Binomial coefficient (see Sect. 7.2) |
| $[r, t]$ | The closed interval, i.e., all real numbers $s$, for which: $r \leq s \leq t$ for $r \leq t$. |
| $]r, t[$ | The open interval, i.e., all real numbers $s$, for which: $r < s < t$ for $r < t$. |
| $d(P, Q)$ | The distance from point $P$ to point $Q$ |
| $\det A$ | The determinant of the matrix $A$ |
| $\square$ | End of proof |

| Symbol | Explanation |
|--------|-------------|
| $(1, 2)$ | The permutation that swaps 1 with 2 |
| $id$ | Identity (identity element in a group of isometries) |
| $\circ$ | Composition symbol for maps of isometries and permutations |
| $|G|$ | Order (number of elements) of the group $G$ (see Definition 2.7) |
| $|g|$ | Order of an element of a group (see Definition 2.19) |
| $+_n$ | Addition modulo $n$ (see p. 21) |
| $\cong$ | Isomorphic to (see Definition 3.22) |
| $\langle a, b \rangle$ | The group generated by $a$ and $b$ (see Definition 2.5) |
| $a \bmod b$ | The remainder when dividing $a$ by $b$ (see p. 21) |
| $U < G$ | $U$ is a subgroup of the group $G$ (see Definition 3.2) |
| $G^+$ | Subgroup of orientation-preserving isometries for a symmetry group $G$ |
| $gH$ | Left coset to $H < G$ and $g \in G$ (see Sect. 3.2) |
| $Hg$ | Right coset to $H < G$ and $g \in G$ (see Sect. 3.2) |
| $[G : U]$ | Index of the subgroup $U$ in $G$ (see Definition 3.15) |
| $\varphi(m)$ | Euler's totient function (see Exercise 5. of Sect. 3.2) |
| $\text{Aut}(G)$ | Automorphism group of $G$ (see Sect. 3.3) |
| $\text{Inn}(G)$ | The group of inner automorphisms of $G$ (see Sect. 3.3) |
| $N \triangleleft G$ | $N$ is a normal subgroup in $G$ (see Definition 3.33) |
| $G/N$ | Factor group of $G$ by $N$ (see Definition 3.35) |
| $G(x)$ | Stabilizer of $x$ in the group $G$ (see Sect. 3.1) |
| $Gx$ | Orbit of $x$ in the group $G$ Ü (see Definition 4.11) |
| $Z(x)$ | Centralizer of an element $x \in G$ (see Sect. 4.4) |
| $G(H)$ | Normalizer of $H < G$ (see Sect. 7.2) |
| $Kx$ | Conjugacy class of an element $x \in G$ (see Sect. 4.3) |

(continued)

| Symbol | Explanation |
|--------|-------------|
| $C(G)$ | Center of the group $G$ (see Exercise 1 of Sect. 3.5) |
| $\Gamma_G$ | Cayley graph of the group $G$ (see Definition 4.48) |
| $\bar{R}$ | Normal closure of $R$ (see page 102) |
| $\langle X \mid R \rangle$ | Presentation of a group with generators $X$ and relations $R$ (see Definition 5.4) |
| $G \times H$ | Direct product of the groups $G, H$ (see Definition 6.2) |
| $G * H$ | Free product of the groups $G, H$ (see Definition 6.9) |
| $N \rtimes H$ | Semidirect product of $N$ and $H$ (see Definition 6.13) |
| $[a, b]$ | The commutator of the group elements $a$ and $b$, defined as $[a, b] = aba^{-1}b^{-1}$ (see Definition 6.4) |
| $[G, G]$ | The commutator subgroup of $G$, also denoted as $G'$ (see Sect. 8.1) |
| $f \preceq g$ | $f(n) \leq ig(jn + k) + ln + m$ (see page 184) |

# Appendix D
# Important Groups

| Group | Explanation | in GAP |
|---|---|---|
| $\{e\}$ | The trivial group | `TrivialGroup()` |
| $\mathcal{E}$ | Symmetry group of the Euclidean plane | |
| $\mathcal{T}$ | Subgroup of translations | |
| $O_n$ | The orthogonal group: Stabilizer of the origin in the symmetry group of $\mathbb{R}^n$ | |
| $SO_n = O_n^+$ | The special orthogonal group: Stabilizer of the origin in the group of orientation-preserving isometries of $\mathbb{R}^n$ | |
| $D_n$ | Dihedral group: Symmetry group of the regular $n$-gon in the Euclidean plane | `DihedralGroup(2n)` |
| $D_\infty$ | The infinite dihedral group (see Example 6.11). | |
| $\mathcal{H}$ | Symmetry group of the hyperbolic plane | |
| $\mathbb{Z}$ | The infinite cyclic group: The integers with the ordinary addition | `FreeGroup(1)` |
| $\mathbb{Z}_n = D_n^+$ | The finite cyclic groups: The integers from 0 to $n-1$ with addition mod $n$ | `CyclicGroup(n)` |
| $\mathbb{Z}_2 \times \mathbb{Z}_2$ | The Klein four-group (see Definition 2.20). | |
| $\mathbb{Z}_n^*$ | The group of prime residue classes mod $n$ (see Exercise 5. of Sect. 3.2). | |
| $S_n$ | Symmetric group over $n$ elements | `SymmetricGroup(n)` |
| $S_X$ | Symmetric group over the set $X$ | |
| $A_n$ | Alternating group over $n$ elements | `AlternatingGroup(n)` |

(continued)

S. Rosebrock, *Visual Group Theory*, Springer Undergraduate Mathematics Series, https://doi.org/10.1007/978-3-662-69365-0

| Group | Explanation | in GAP |
|-------|-------------|--------|
| $F_n$ | The free group of rank $n$ | `FreeGroup(n)` |
| $G_{(n,m)}$ | Symmetry group for the decomposition into regular $n$-gons, where always $m$ meet at a vertex (see p. 88) | |
| $G_{m,n}$ | The group presented by $\langle x, y \mid x^m, y^n, xy = y^2x \rangle$ | |
| $Q$ | The Quaternion group (see Exercise 6. of Sect. 5.1) | `SmallGroup(8,4)` |
| $GL(n, \mathbb{R})$ | The linear group (invertible linear mappings of $\mathbb{R}^n$ onto itself) | |

# Appendix E
# Used GAP Commands

| GAP Command | Explanation |
|---|---|
| ActionHomomorphism(G,M) | Generates a homomorphism from $G$ to the symmetric group over $M$ |
| AllGroups(n) | Returns all groups of order $n$ |
| AlternatingGroup(n) | The alternating group over $n$ elements |
| AssignGeneratorVariables(F) | Assigns names to generators |
| AutomorphismGroup(G) | The automorphism group of the group $G$ |
| Binomial(n,k) | The binomial coefficient $\binom{n}{k}$ |
| Centralizer(G,g) | The centralizer of $g$ in the group $G$ |
| Centre(G) | The center of the group $G$ |
| ConjugacyClasses(G) | The conjugacy classes of the group $G$ |
| ConjugacyClassesSubgroups(G) | Conjugacy classes of subgroups |
| CyclicGroup(n) | The cyclic group of order $n$ |
| DerivedSubgroup(G) | Commutator subgroup of $G$ |
| DihedralGroup(n) | The dihedral group of order $n$ |
| DirectProduct(G,H) | The direct product of the groups $G$ and $H$ |
| Elements(G) | The elements of the group $G$ |
| FactorGroup(G,N) | The factor group of $G$ modulo the normal subgroup $N$ |
| FreeGroup("x", "y") | The free group generated by $x$ and $y$ |
| Group(g,h) | The group generated by $g, h$ |
| GroupHomomorphismByImages | A group homomorphism by specifying pre-images and images |
| GroupSatisfiesC(G,8) | Checks whether C(8) is satisfied |
| GroupSatisfiesT(G,3) | checks whether T(3) is satisfied |

(continued)

| GAP command | Explanation |
|---|---|
| Image(hom) | The image of the homomorphism *hom* |
| Index(G,H) | The index of *H* in *G* |
| IsCyclic(G) | Checks whether *G* is a cyclic group |
| IsNilpotentGroup(G) | Checks whether *G* is nilpotent |
| IsNormal(G,H) | Checks whether *H* is normal in *G* |
| IsSolvable(G) | Checks whether *G* is solvable |
| IsomorphismGroups(G,H) | An isomorphism between *G* and *H* |
| Kernel(hom) | The kernel of the homomorphism *hom* |
| List(l,f) | Applies the function *f* to the elements of the list *l* |
| MinimalGeneratingSet(G) | A smallest generating system for the group *G* |
| MultiplicationTable(Elements(G)) | Multiplication table for the group *G* |
| NaturalHomomorphismByNormalSubgroup(G,N) | Forms the homomorphism $G \to G/N$ |
| NilpotencyClassOfGroup(G) | Nilpotency class of the group *G* |
| Normalizer(G,H) | The normalizer of *H* in *G* |
| NormalSubgroups(G) | The normal subgroups of the group *G* |
| Orbit(G,g) | The orbit of *g* in the group *G* |
| Order(G) | The order of the group *G* |
| Order(g) | The order of the element *g* |
| PiecesOfGroup(G) | List of Pieces |
| PresentationFpGroup(G) | A presentation for the group *G* |
| SemidirectProduct(H,hom,N) | Semidirect product of *N* with *H* |
| SimplifyPresentation(P) | Simplifies a group presentation *P* with Tietze transformations |
| SmallGroupsInformation(n) | The groups of order *n* |
| Stabilizer(G,g) | The stabilizer of *g* in the group *G* |
| StructureDescription(G) | Gives the standard name of the group *G* |
| Subgroup(G,[g,h,j]) | The subgroup of the group *G* generated by *g*, *h* and *j* |
| SylowSubgroup(G,n) | A Sylow subgroup of the group *G* |
| SymmetricGroup(n) | The symmetric group over *n* elements |
| TzPrintRelators(P) | The relators of the presentation *P* |
| TzSubstitute(P) | Simplifies the presentation *P* using Tietze transformations |

# Appendix F
# Hints for the Exercises

## Section 1.1
1. There are reflections with three mirror axes through each triangle midpoint and six mirror axes through the vertices of the decomposition. Rotations can also be performed at these points. There are further rotations of 180° around the midpoints of the edges. Translations exist in various directions, which can be composed of translations in only two directions. Similarly, there are glide reflections, which can be composed of reflections and translations.
2. There are $n$ reflections and $n$ rotations by $2\pi/n$ and multiples thereof.
3. See for example Fig. 6.2.

## Section 1.2
1. A rectangle allows 2 reflections and rotations of 180 and 0°. A parallelogram, apart from the identity, can only be rotated by 180° at the midpoint.
2. A rotation of 180°. In this case, the commutative law applies to the mirror axes.
3. See for example Fig. 2.3.

## Section 1.3
1. $(1, 3)(2, 4)$ yes (the 180° rotation), $(1, 2, 3)$ no.
2. 2 mirror axes parallel to each other to generate the translation and the third perpendicular to them.
3.

| $\circ$ | $id$ | $d_{180}$ | $s_a$ | $s_b$ |
|---------|------|-----------|-------|-------|
| $id$ | $id$ | $d_{180}$ | $s_a$ | $s_b$ |
| $d_{180}$ | $d_{180}$ | $id$ | $s_b$ | $s_a$ |
| $s_a$ | $s_a$ | $s_b$ | $id$ | $d_{180}$ |
| $s_b$ | $s_b$ | $s_a$ | $d_{180}$ | $id$ |

S. Rosebrock, *Visual Group Theory*, Springer Undergraduate Mathematics Series, https://doi.org/10.1007/978-3-662-69365-0

**Section 1.4**

1. The sphere allows reflections through all planes passing through its center, reflections and rotations around all axes through the center and a point reflection. The isometries of the tetrahedron are described in Sect. 2.5.
2. $(1, 7)(2, 3, 4, 8, 5, 6)$, if you rotate first.
3. (a) With a translation, we move any point to its image point. We rotate another $n - 1$ linearly independent points onto each other and may need to mirror once at the end.
   (b) We represent translation and rotation each as a product of 2 reflections.

**Section 2.1**

1. The elements of $D_4$ are

   $$\{id, (1, 2, 3, 4), (1, 3)(2, 4), (1, 4, 3, 2), (1, 3), (2, 4), (1, 4)(2, 3), (1, 2)(3, 4)\}.$$

   It holds that $(1, 2, 3, 4)^{-1} = (1, 4, 3, 2)$, and the other elements are inverse to themselves.
2. $0$ is the identity element. For $p \in \mathbb{Q}$, $-p$ is the inverse element.
3. The identity element is $e(x) = 0 \cdot x + 0$. The inverse to $f(x) = ax + b$ is $g(x) = -ax - b$.
4. There are of course unlimited possibilities. For example, $a \diamond b = a - 2ab + b$ is associative and commutative. There is an identity element, namely $0$. The element $1$ is inverse to itself, but $3$ has no inverse. Therefore, it is not a group.
5. All rotations are obtained through $d, d^2, \ldots, d^n = id$ and all reflections through $ds, d^2s, \ldots, d^n s = s$.

**Section 2.2**

1. Yes.
2. $(0, 0)$ forms the identity element. The inverse to $(a, b)$ is $(-a, -b)$. The group is commutative because: $(a, b) + (n, m) = (a + n, b + m) = (n, m) + (a, b)$.
3. $(1, 2, 3, \ldots, n)$, $(1, 2, 3, \ldots, n)^2 = (1, 3, 5, \ldots, n - 1)(2, 4, 6, \ldots, n)$ for even $n$ and $(1, 2, 3, \ldots, n)^2 = (1, 3, 5, \ldots, n, 2, 4, 6, \ldots, n - 1)$ for odd $n$, etc.
4. The odd numbers do not form a group, because there is no odd number $u$ such that $k + u = k$ for odd numbers $k \in \mathbb{Z}$. Therefore, no identity element can be found.
5. The group $D_n$ is not commutative for $n \geq 3$. A reflection cannot be interchanged with a rotation by an angle different from $180°$.
6. Every integer can be written as a sum of ones or minus ones, i.e. $\mathbb{Z} = \langle 1 \rangle$. From 2 and 3, we can generate 1 by writing $3 + (-2)$. We already know that 1 generates $\mathbb{Z}$, so $\mathbb{Z} = \langle 2, 3 \rangle$. Any set of coprime integers generates $\mathbb{Z}$.

**Section 2.3**

1. 5 and 4.
2. Yes. If you add multiples of 4 mod 12 to each other, you get another multiple of 4 mod 12. The inverse of 4 is 8.
3. Yes. Each element is inverse to itself. For example, $3 \cdot 5 = 7$.

4. The length of a translation vector behaves like a real number under addition.
5. (a) 1 must be the identity element, and then 0 has no inverse.
   (b) In $\mathbb{J}_8$ the element 2 has no inverse. If $n$ is a prime number, then $\mathbb{J}_n$ is a group with respect to multiplication mod $n$.
6. The elements 1, 5, 7, 11 each generate $\mathbb{Z}_{12}$ on their own. For 5, this is seen as follows: $5, 5 +_{12} 5 = 10, 5 +_{12} 10 = 3, 5 +_{12} 3 = 8, 5 +_{12} 8 = 1$, and since 1 generates $\mathbb{Z}_{12}$, we know that 5 also generates $\mathbb{Z}_{12}$.

## Section 2.4
1. $d = bc^{-1}$.
2. After multiplying both sides of the equation from the left by $(1, 3)$ we get: $x = (1, 2, 3, 4)$.
3. According to item 3 of Theorem 2.16 it holds that $(g^{-1})^{-1} = g$, and thus $((g^{-1})^{-1})^{-1} = g^{-1}$.

## Section 2.5
1. The order of 3 in $\mathbb{Z}_7$ is 7.
2. Infinite.
3. In the regular 12-gon, choose any reflection and the rotations by 120 and 90°.
4. Let $a$ and $b$ be two arbitrary elements of the group. Since $a, b$ have order 2, it follows that $a = a^{-1}$ and $b = b^{-1}$. $ab$ has order 2, so $abab = 1$. This implies $ab = b^{-1}a^{-1} = ba$, which was the claim.
5. For example: $(1, 3) \circ (2, 4) = (1, 3)(2, 4)$ and $(1, 2) \circ (1, 3) = (1, 2, 3)$.
6. There are uncountably many angles between 0 and 360°, and thus uncountably many rotations that map the circle onto itself.
7. 48. Indeed, an octahedron can be inscribed in a cube so that each vertex of the octahedron lies in the center of a side of the cube. Every isometry of the cube is then one of the octahedron and vice versa. The symmetry groups of the cube and the octahedron are therefore isomorphic.
8. The multiplication table of the Klein four-group (see Definition 2.20) is:

| $\circ$ | $id$ | $d$ | $s_a$ | $s_b$ |
|---|---|---|---|---|
| $id$ | $id$ | $d$ | $s_a$ | $s_b$ |
| $d$ | $d$ | $id$ | $s_b$ | $s_a$ |
| $s_a$ | $s_a$ | $s_b$ | $id$ | $d$ |
| $s_b$ | $s_b$ | $s_a$ | $d$ | $id$ |

The group is abelian if and only if the table is symmetric along the main diagonal from upper left to lower right.
9. If $\mathbb{Q}$ were finitely generated, and $\{a_1/b_1, \ldots, a_n/b_n\}$ were the generators, then every element of $\mathbb{Q}$ would have to be represented as $c/d$, where $d$ is the least common multiple of the $b_i$, which is not true.
10. Assume $g^n = 1$. Then $g^n g^{-n} = 1$ and $(g^{-1})^n = g^{-n}$, which implies $(g^{-1})^n = 1$.

11. Let $p$ be the order of $g$ and $q$ the order of $h$. It holds that $(gh)^{pq} = g^{pq}h^{pq} = e$, because $g$ and $h$ commute. If $n < pq$, then $g^n \neq e$ or $h^n \neq e$. Also, $g^n \neq h^{-n}$.

## Section 3.1

1. Every rational number is also a real number. The product of two rational numbers is rational.
2. The product of two elements from the intersection is again in the intersection.
3. No, because you can generate a translation with 2 rotations about different points of the plane.
4. If $c$ and $d$ are two elements of $G$, we write them as products in $a$ and $b$ and their inverses. If you multiply the two products, you can slide one past the other because $ab = ba$, and the product remains the same. It follows that $cd = dc$, and $G$ is abelian.
5. If $a$ has order $n$ and $b$ has order $m$, then $ab$ has at most order $nm$, and $a^{-1}$ also has order $m$.
6. The trivial group, $\mathbb{Z}_n$ for $n \in \mathbb{N}$ ($n > 1$) and $\mathbb{Z}$.
7. If for any two elements $g, h$ of a group it is clear what $g \cdot h$ is, then the group is completely determined. But there are only finitely many possibilities for the result of the product $g \cdot h$.
8. () has order 1,

   $(1, 10)(2, 11)(3, 12)(4, 13)(5, 14)(6, 15)(7, 16)(8, 17)(9, 18)$ has order 2,
   $(1, 13, 7)(2, 14, 8)(3, 15, 9)(4, 16, 10)(5, 17, 11)(6, 18, 12)$ has order 3,
   $(1, 16, 13, 10, 7, 4)(2, 17, 14, 11, 8, 5)(3, 18, 15, 12, 9, 6)$ has order 6, and
   $(1, 11, 3, 13, 5, 15, 7, 17, 9)(2, 12, 4, 14, 6, 16, 8, 18, 10)$ has order 9.

9. Groups with prime order are cyclic. Let $G$ be a group of order 4. If an element of $G$ has order 4, then $G$ is cyclic. Otherwise, $G$ is the Klein four-group.

## Section 3.2

1. A regular $n$-gon can be inscribed in a regular $2n$-gon. $[D_{2n} : D_n] = 2$.
2. The sum of two elements from $n\mathbb{Z}$ is, due to $n \cdot k + n \cdot m = n \cdot (k+m)$, again in $n\mathbb{Z}$. The identity element 0 is in $n\mathbb{Z}$, and the inverse to $n \cdot k$ is $n \cdot (-k)$. $[\mathbb{Z} : n\mathbb{Z}] = n$.
3. Generate the cube group in GAP as described at the end of Sect. 2.1, and then:

   ```
   gap> S4:=Subgroup(W,[(1,3)(5,7), (6,3)(5,4), (1,8)(2,7)]);
   Group([ (1,3)(5,7), (3,6)(4,5), (1,8)(2,7) ])
   gap> Order(S4);
   24
   gap> Index(W, S4);
   2
   ```

4. Each left coset has the form $r + \mathbb{Z}$ for $0 \leq r < 1$. Therefore, there are uncountably many cosets.
5. $\mathbb{Z}_m^*$ has order $\varphi(m)$. Therefore, according to Corollary 3.21, for all elements $a \in \mathbb{Z}_m^*$, $a^{\varphi(m)} = id$.
6. $U$ is isomorphic to the Klein four-group. $[D_6 : U] = 3$.

7. From $g_1 H = g_2 H$ it follows by multiplication of $g_1^{-1}$ on both sides, that $g_1^{-1} g_2 H$ must lie in $H$ and from this the assertion. Conversely, if $g_1 H \neq g_2 H$, there must be an element $x \in g_1 H$ that is not in $g_2 H$. $g_1^{-1} x \in H$, but $g_1^{-1} x \notin g_1^{-1} g_2 H$.

8. Let $g \in U \cap V$. If $n = |U|$ and $m = |V|$, then $g^n = g^m = id$. If $n$ and $m$ are coprime, one can consider with the help of the Euclidean algorithm that $g = id$ must hold.

## Section 3.3

1. This follows directly from the following statement: If $u \neq v \in U$ then $gug^{-1} \neq gvg^{-1}$.

2. If $x$ has order $n$, then it follows from the homomorphism property that $1 = \phi(1) = \phi(x^n) = \phi(x)^n$. So $\phi(x)$ has at most the order $n$. If $\phi(x)$ had order $k < n$, then with $1 = \phi(x)^k = \phi(x^k)$ it would follow that $x$ has order $k$.

3. Every isometry that maps a circle about the origin onto itself is in $O_2$ and vice versa.

4. The mapping $\phi : \mathbb{Z} \to n\mathbb{Z}$ given by $k \to n \cdot k$ is an isomorphism.

5. $\phi_n(k + m) = n \cdot (k + m) = nk + nm = \phi_n(k) + \phi_n(m)$.
   $\psi_n(k + m) = n + k + m \neq (n + k) + (n + m) = \psi_n(k) + \psi_n(m)$.

6. The automorphism group of $\mathbb{Z}$ is $\mathbb{Z}_2$, because 1 can only be mapped to 1 or to $-1$. A homomorphism $\phi : \mathbb{Z} \to \mathbb{Z}$ with $\phi(1) = n$ and $n > 1$ or $n < -1$ is not surjective.

7. If $n$ is prime, then 1 can be mapped to any element of $\mathbb{Z}_n$ except 0, and the automorphism group is isomorphic to $\mathbb{Z}_{n-1}$. What happens for non-prime $n$? For example:

```
gap> au:=AutomorphismGroup(CyclicGroup(12));;
gap> Order(au); IsCyclic(au);
4
false
```

8. Let $n = k \cdot m$. The elements $0, k, 2k, 3k, \ldots, (m-1)k$ form a subgroup of $\mathbb{Z}_n$ which is isomorphic to $\mathbb{Z}_m$.

## Section 3.4

1. We map the real number $r$ to a rotation by $r$ degrees. The kernel contains all rotations by multiples of $360°$. It is thus isomorphic to $\mathbb{Z}$.

2. $U$ consists of the reflection and the identity. If $U = \langle (1, 2) \rangle$, then we obtain the coset $(1, 2, 3)U = \{(1, 2, 3), (1, 3)\}$. $U$ is not normal in $G$, which can be seen by calculating $U(1, 2, 3)$.

3. Since $H$ has index 2, we can write $G = H \cup gH$ and similarly $G = H \cup Hg$. This implies $gH = Hg$.

4. According to Exercise 2. of Sect. 3.1, $H \cap H'$ is a subgroup. Let $h \in H \cap H'$. It follows that $hG = Gh$, because $h \in H'$, and $H'$ is normal in $G$.

5. The cosets $g \circ D_2$ and $D_2 \circ g$ are different for $g = (1, 4, 7, 2, 5, 8, 3, 6)$.

6. $p$ commutes with all elements of $D_6$. Therefore, $U$ is normal in $D_6$. If $d \in D_6$ is a rotation by $60°$ and $s \in D_6$ is a reflection, then we have the following cosets:

$U, d \circ U, d^2 \circ U, s \circ U, ds \circ U, d^2 s \circ U$. These cosets behave like the elements of the group $D_3$.

7. We have $|U| = 3$ and $[D_6 : U] = 4$. Let $s, s', s'' \in D_6$ be the reflections along lines through opposite vertices and $t, t', t'' \in D_6$ through opposite midpoints of edges in a regular hexagon. Then

$$U = \{id, d^2, d^4\}, \quad dU = Ud = \{d, d^3, d^5\}$$

and

$$sU = Us = \{s, s', s''\}, \quad tU = Ut = \{t, t', t''\}.$$

$(sU)^2 = (tU)^2 = (dU)^2 = U$. $dU \circ sU = tU$. $tU \circ sU = dU$. The homomorphism $\phi \colon D_6 \to D_6/U$ is given by: $g \in D_6$ is mapped to $gU$.

8. Let $\phi_h(g) = hgh^{-1}$ be an inner automorphism of $G$. Then for all $\psi \in \mathrm{Aut}(G)$ we have

$$\forall g \in G, \quad \psi \circ \phi_h \circ \psi^{-1}(g) = \phi_{\psi(h)}(g).$$

## Section 3.5

1.(a) Let $b, c \in C(G)$. Then for any $g \in G$: $bcg = bgc = gbc$, and thus $bc \in C(G)$.

(b) The rotation by $180°$ is in the center of $D_4$.

(c) If $D_n$ has a rotation by $180°$, then this is in the center. Otherwise, the center is trivial.

(d) It holds that $S_3 \cong D_3$, and $D_3$ has a trivial center according to (c).

2. Let $G$ be the symmetry group of $\mathbb{R}^n$ and $O_n$ the stabilizer of the origin. We define a mapping $\phi \colon G \to O_n$: Let $g \in G$ and $P$ the image of the origin under $g$. Let $\tau$ be the translation that maps $P$ to the origin. Then let $\phi(g) = \tau \circ g$.

3. On page 55, it is described how to find $\tau$ and $d'$.

## Section 4.1

1. Up to isomorphism, there is only one group of order 3. Since both groups have order 3, they must be isomorphic.

2. For every pair of vertices $i, j$ of the $n$-simplex, there is an edge that connects $i$ with $j$. The hyperplane through the midpoint of this edge, which contains all other vertices, maps the $n$-simplex onto itself through a reflection. Therefore, the symmetry group of the $n$-simplex contains all transpositions and thus the entire group $S_n$ according to Theorem 4.2.

3. See the proof of Theorem 5.19.

## Section 4.2

1.(a) $G(\text{Vertex}) \cong D_3$.

(b) $G(\text{Edge}) \cong D_2$.

(c) $G(\text{Face}) \cong D_4$.

(d) $D_4 \times \mathbb{Z}_2$.

(e) $\mathbb{Z}_2 \times \mathbb{Z}_2 \times \mathbb{Z}_2$, the group of a cuboid without square faces.

(f) This depends on the position of the face in relation to the pair of edges.

2. (a) No.

(b) Yes. A square can be drawn into an octagon by taking every second vertex of the octagon as a vertex of the square.

(c) Yes. Analogous to (b), a heptagon can be embedded into a 56-gon.

(d) No.

(e) Yes: The stabilizer of a vertex in the group $S_4$ of the tetrahedron is isomorphic to the group $D_3$.

3. We use the vertex designations of the cube of Fig. 1.8:

```
gap> a:=(1,2)(5,6)(4,3)(8,7);;
gap> b:=(1,3)(5,7);;c:=(5,4)(6,3);;
gap> W:=Group(a,b,c);;
gap> diag := Orbit( W, [1,7], OnSets );
[ [ 1, 7 ], [ 4, 6 ], [ 2, 8 ], [ 3, 5 ] ]
gap> hom := ActionHomomorphism( W, diag, OnSets );;
gap> H := Image( hom );
Group([ (1,2)(3,4), (1,3)(2,4), (2,4), (1,4,2,3), (2,4,3) ])
gap> s4 := SymmetricGroup( 4 );;
gap> IsomorphismGroups( H, s4 );
[ (1,2)(3,4), (1,3)(2,4), (2,4), (1,4,2,3), (2,4,3) ] ->
[ (1,2)(3,4), (1,3)(2,4), (1,3), (1,3,2,4), (1,2,3) ]
gap> k := Kernel( hom );
Group([ (1,7)(2,8)(3,5)(4,6) ])
```

The orbit of a diagonal consists of all diagonals. We obtain a homomorphism by considering the action of a group element on the diagonals; the cube group operates on the diagonals. The image is isomorphic to the group $S_4$ with ker(*hom*) equal to the orientation-reversing reflection $s_M$ through the center of the cube.

4. By a rotation, we can map any two given vertices of the cube onto each other. This is also possible with the edges and the faces. All three operations are therefore transitive.

## Section 4.3

1. If $g$ is conjugate to $h$, then there is a $t \in G$ with $t^{-1}gt = h$. If $h$ is conjugate to $j$, then there is an $s \in G$ with $s^{-1}hs = j$. It follows that $j = s^{-1}hs = s^{-1}t^{-1}gts$.

2. $t^{12} + t^{10} + t^7 + 2t^6 + t^5 + 2t^4 + 2t^3 + 3t^2 + t$.

3. $f_g(h)f_g(h') = g^{-1}hgg^{-1}h'g = f_g(hh')$.

4. These are rotation groups isomorphic to $H$ around other points of the plane.

5. Stabilizers of the other cube faces.

6. Conjugate subgroups to $\{id, t\}$ consist of the identity and another reflection of $D_4$.

## Section 4.4

1. Let $s$ be a face of the cube. Then $|W^+| = |W^+(s)| \cdot |W^+s| = 4 \cdot 6 = 24$.

2. The centralizer of a group element is the set of all elements that commute with it. In an abelian group, every element commutes with a given one, and therefore

**Fig. F.1**   Cayley graph for
the group $D_2$

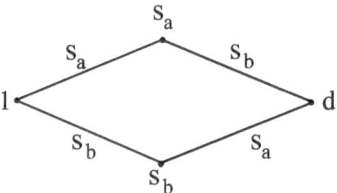

the centralizer is the whole group. Only rotations commute with rotations, no
reflections. An exception is the rotation by $180°$, which commutes with all group
elements.

3. The conjugacy classes are $\{id\}$, $\{(2, 4), (1, 3)\}$, $\{(1, 2)(3, 4), (1, 4)(2, 3)\}$,
   $\{(1, 2, 3, 4), (1, 4, 3, 2)\}$ and $\{(1, 3)(2, 4)\}$.
4. Let $h \in G(x)$. Then it follows that $ghg^{-1}(y) = gh(x) = g(x) = y$.
5. If $x \in S_r^{n-1}$ and $g \in O_n$, then $g(x) \in S_r^{n-1}$. For $x, y \in S_r^{n-1}$ there is a reflection
   $s \in O_n$ with $s(x) = y$.
6. An element $g \in G$ is in the center of a group if and only if it commutes with all
   group elements. This is the case precisely when it lies in the centralizer of every
   group element.

## Section 4.5

1. For the group $D_5$, a "double pentagon" results, similar to the double triangle of
   Fig. 4.9.
2. See Fig. F.1.
3. For two vertices of the Cayley graph, i.e., group elements $g, h \in G$, there is a
   group element, namely $v = hg^{-1}$, which transforms $g$ to $h$. Subsequent edges
   are transferred by the same group element.
4. If $g, h \in G$ are two vertices of the Cayley graph, then write the element $hg^{-1}$ in
   the generators. This word can be read off the edges of a path from $g$ to $h$.
5. A lattice of squares results.
6. Let $g, h \in G$ with $d_\Gamma(g, h) > |G|$. This means the word $g^{-1}h$ has in the
   generators $X$ and their inverses length $n > |G|$, so $g^{-1}h = a_1 \ldots a_n$ with
   $a_i \in X^{\pm 1}$, and $n$ is minimal. The elements $v_i = a_1 \ldots a_i$ for $i = 1, \ldots n$ cannot
   all be different. From $v_k = v_j$ for $k < j$ it follows that $a_{k+1} \ldots a_j = 1$, and
   $g^{-1}h$ has a shorter representation, in contradiction to the minimality of $n$.
7. The set of vertices is $\mathbb{Z}$. There are edges labeled with 2 from $i$ to $i + 2$ and edges
   labeled with 3 from $i$ to $i + 3$ for all $i \in \mathbb{Z}$.

## Section 4.6

2. The fundamental region is a right-angled triangle, one eighth of a square with
   vertices in the middle of a side, an adjacent vertex of the square and the center of
   the square (see Fig. 2.1).
3. $G_{(4,4)}/G \cong D_4$.
4. Each hexagon is divided by mirror axes into 12 right-angled triangles, one of
   which serves as a fundamental region. Each of these triangles has the same

interior angles as the triangles of Fig. 4.12, so that the reflections over the sides
of such a triangle generate the same group.

5. The Cayley graph is a hexagon.

6. $H$ is not normal in $S_4$, which can be seen from the following GAP code:

```
gap> S4:=Group((1,2,3,4),(1,2));
Group([ (1,2,3,4), (1,2) ])
gap> H:=Subgroup(S4,[(1,2,4,3)]);;
gap> h:=Elements(H);
[ (), (1,2,4,3), (1,3,4,2), (1,4)(2,3) ]
gap> (1,2)*h;
[ (1,2), (1,4,3), (2,3,4), (1,3,2,4) ]
gap> h*(1,2);
[ (1,2), (2,4,3), (1,3,4), (1,4,2,3) ]
```

## Section 5.1

1. Since $a = a^{-1}, b = b^{-1}, aba = b$ and $bab = a$, only the 4 elements $id, a, b, ab$
   remain as group elements. This can also be seen from the Cayley graph in Fig. F.1
   of the solution of Exercise 2. of Section 4.5.

2. For example: $\langle a, b \mid a^2, b^2, (ab)^2 \rangle$, $\langle a, b \mid a^2, ab^2a, (ab)^2 \rangle$.

3. We remove the generator $\tau$ and the relation $s^2 = \tau$. The last relation becomes
   redundant, and we get $\langle s \mid \ \rangle$ as a presentation of $\mathbb{Z}$.

4. We start with the vertices $id, d, d^2, d^3, d^4$ and $s, ds, d^2s, d^3s, d^4s$. We draw an
   oriented edge from $d^k$ to $d^{k+1}$ for $0 \le k \le 3$ and because of the relation $d^5$ from
   $d^4$ to $id$ . All these edges are labeled with $d$. An unoriented edge labeled with $s$
   connects $d^k$ with $d^ks$ for $0 \le k \le 4$. In addition, because of the relation $sdsd$,
   there are oriented edges labeled with $d$ from $d^{k+1}s$ to $d^ks$ for $0 \le k \le 3$ and
   from $s$ to $d^4s$.

5. For example: $\langle x \mid x^n \rangle$.

6. (a) The elements $\{\pm 1, \pm k, \pm i, \pm j\}$ are all different, and there are no more, as
   can be seen from the relations.

   (b) The element $-1$. The elements $\pm k, \pm i, \pm j$ all have order 4, and commutators
   between them are $-1$.

   (c) The group $D_4$ contains 5 elements of order 2, the group $Q$ according to (b)
   only one.

   (d) For example: $U = \langle k \rangle = \{1, k, -1, -k\}$. Convince yourself that $iU = Ui, jU = Uj$.

   (e) Set $x = i$ and $y = j$.

## Section 5.2

1. In Fig. F.2 a part of the Cayley graph of the free group generated by $a$ and $b$ is
   shown, where all horizontal edges are labeled with $a$ and all vertical edges with
   $b$. It is a tree, where each *vertex degree* (number of outgoing edges of a vertex)
   is 4.

2. Show that the coset $aH$ contains all elements of odd length of $F_2$.

3. In a free group, a non-trivial group element commutes only with powers of itself.

**Fig. F.2**  Cayley graph for
the free group of rank 2

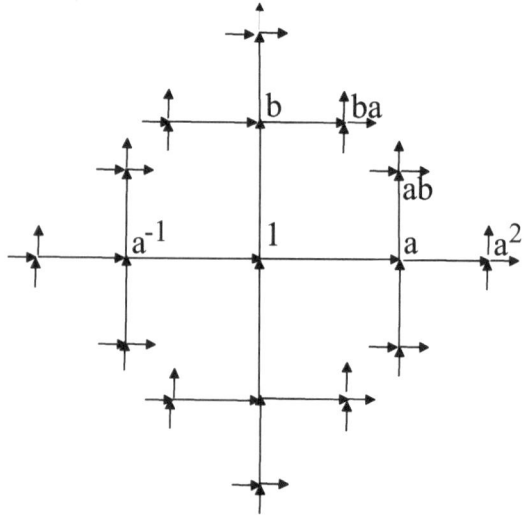

4. Here is the corresponding GAP-code:

```
gap> F := FreeGroup( "a", "b");
<free group on the generators [ a, b ]>
gap> AssignGeneratorVariables(F);
#I  Assigned the global variables [ a, b ]
gap> G:=F/[a^4*b^-2,a*b*a*b^-1];
<fp group on the generators [ a, b ]>
gap> Order(G);
16
gap> Elements(G);
[ <identity ...>, a, b, a^2, b^2, a*b, a^3, a*b^2, a^2*b^-1,
    b^-1, a^-2, a^-1*b, a*b^-1, a^-1, a^2*b, a^-1*b^-1 ]
```

## Section 5.3

1. First, sort the generators $g_1, \ldots, g_n$ of the abelian group. Each word can then be
   brought into the form: $g_1^{e_1} \cdots g_n^{e_n}$ with $e_i \in \mathbb{Z}$.
   In the Cayley graph, a given word can be realized as a path starting at 1. The
   word is trivial if and only if the path ends at the vertex 1.
2. Inverting a relation: Let $G = \langle X \mid R \rangle$ and $r \in R$. Since $r^{-1} \in \bar{R}$, we can form the
   presentation $\langle X \mid R, r^{-1} \rangle$ using (5.3). Subsequently, we eliminate the relation $r$
   using $(5.3)^{-1}$.
3. The easiest way to demonstrate this is with GAP:

```
gap> F:=FreeGroup("x","y");;
gap> AssignGeneratorVariables(F);
#I  Assigned the global variables [ x, y ]
gap> G:=F/[x*y*y*x^-1*y^-3, y*x*x*y^-1*x^-3];
<fp group on the generators [ x, y ]>
gap> Order(G);
1
```

However, it is not immediately clear why this is the case. Try proving it with a clever sequence of Tietze transformations.

4. $\langle a, b, s \mid ab = ba, s^2, as = sb \rangle$ with translations $a, b$ and the reflection $s$. A normal form is $a^k b^n s^\delta$ with $\delta \in \{0, 1\}$ and $n, k \in \mathbb{Z}$. In fact, each subword $s b^d a^e s$ can be transformed into $a^d b^e$, so that all $s$ (except at most one) can be eliminated.

5. Two words are conjugate in $F$ if and only if, after free reduction, one can be found as a subword in the other, such that the complement can be freely reduced to 1.

## Section 6.1

1. (a) $\langle a, b, c \mid (ab)^4, (bc)^4, (ac)^2 \rangle$.

   (b) The square grid can be shifted to the right (or left) and upwards (or downwards). These shifts each generate a $\mathbb{Z}$-component. The shifts commute with each other.

2. The order is the least common multiple of $m$ and $n$.

3. The group is generated by a translation $\tau$ and a reflection $s$ that commute with each other. So $\langle s, \tau \mid s^2, s\tau = \tau s \rangle$, and according to Theorem 6.5 this is a presentation of $\mathbb{Z} \times \mathbb{Z}_2$.

4. $\psi(kp + mq) = (m, k)$. That $\mathbb{Z}_n$ is isomorphic to $\mathbb{Z}_p \times \mathbb{Z}_q$ ($p$ and $q$ coprime) can also be seen from the fact that both are cyclic groups of the same order.

5. (a) The associative law is violated: $(4 \diamond 2) \diamond 2 = 2 \diamond 2 = 1$, but $4 \diamond (2 \diamond 2) = 4 \diamond 1 = 4$.

   (b) $M$ should only consist of the numbers where each of the prime factors $2, 3, 5$ and $7$ appears at most once.

   (c) $\mathbb{Z}_2 \times \mathbb{Z}_2 \times \mathbb{Z}_2 \times \mathbb{Z}_2$.

6. The element $(e, 1)$ lies in the center of $A_n \times \mathbb{Z}_2$, where $e \in A_n$ is the identity element.

## Section 6.2

1. Let $G = U * V$. If $1 \neq u \in U$ and $1 \neq v \in V$, then $i(u)j(v)$ has infinite order in $G$, where $i: U \to G$ and $j: V \to G$ are the inclusions.

2. According to the definition of the free group, $\langle a, b \mid \ \rangle$ is a presentation of the free group of rank 2. Since $\mathbb{Z} = \langle t \mid \ \rangle$, the assertion follows.

## Section 6.3

1. $mn$.

2. Let $G_P < \mathcal{E}(l)$ be the subgroup that leaves $P$ fixed, where $P$ is a point in $l$. The translation subgroup $T < \mathcal{E}(l)$ is normal in $\mathcal{E}(l)$. From Theorem 6.17, it follows that $\mathcal{E}(l) = T \rtimes G_P$.

## Section 6.4

1. If the orbit of $P$ has only finitely many points in each disk $D \subset \mathbb{R}^2$, then one can choose $D$ as a disk which contains $P$. Since $D$ only contains finitely many points of the orbit, one of these points has the smallest possible distance $l$ to $P$. The disk $D'$ with radius $l/2$ and center $P$ contains no image of $P$ under $G$ except $P$ itself.

2. $\langle a, b \mid ab = ba \rangle$.

## Section 7.1

1. Let $n = 2^k$. From the relation $xy = y^2x$ it follows that $xy^{2^{k-1}}x^{-1} = y^{2^k} = y^n = 1$ and therefore $y^{2^{k-1}} = 1$. Now $xy^{2^{k-2}}x^{-1} = y^{2^{k-1}} = 1$ and therefore $y^{2^{k-2}} = 1$ etc. until we have $y = 1$.

2. Squaring the relation $y^2 = xyx^{-1}$ gives $y^4 = xy^2x^{-1}$. Replacing $y^2$ in it by $xyx^{-1}$, one obtains $y^4 = x^2yx^{-2}$. Now it follows that

$$y^2 = y^8 = xy^4x^{-1} = x^3yx^{-3} = y$$

   and therefore $y = 1$.

3. $G_{2,3} = \langle x, y \mid x^2, y^3, xy = y^2x \rangle$. The third relation can be transformed by Tietze transformations into $xy = y^{-1}x^{-1}$ and then to $(xy)^2$. The presentation one obtains corresponds to the one of Theorem 5.5.

## Section 7.2

1. $G(G) = G$.
2. $\mathbb{Z}_{15}$ is abelian, i.e. $\mathbb{Z}_{15}(H) = \mathbb{Z}_{15}$.
3. For example with GAP:

```
gap> G:=Group((1,3),(1,2,3,4));
Group([ (1,3), (1,2,3,4) ])
gap> Normalizer(G,Subgroup(G,[(2,4)]));
Group([ (2,4), (1,3) ])
gap> Elements(last);
[ (), (2,4), (1,3), (1,3)(2,4) ]
```

   But one can also easily deduce that besides $s$ the normalizer also contains the reflection over the axis perpendicular to $s$ and the rotation by $180°$.

## Section 7.3

1. According to Theorem 7.12, all groups of order $p^2$ are abelian. If such a group contains an element of order $p^2$, it is the group $\mathbb{Z}_{p^2}$. Otherwise, all nontrivial elements have order $p$. Let $x$, $y$ be two such elements, where $x$ is not a power of $y$. Since the group is abelian, these two elements must commute, and we obtain the presentation $\langle x, y \mid x^p, y^p, yx = xy \rangle$.

2. $A_4 \not\cong D_6$, because the dihedral group contains rotations of order 6, but in the group $A_4$ all nontrivial elements have orders 2 and 3.

3. This follows directly from Theorem 7.13.

4. If the group $A_5$ had a subgroup of order 30, then according to Exercise 3. of Sect. 3.4, because it has index 2 in $A_5$, it would be a normal subgroup. However, according to Theorem 4.41, the group $A_5$ is simple.

5. All permutations that can be represented in the form (a,b,c,d). There are 30 different ones.

6. (a) If $a$ and $b$ are two elements of the group $S_n$, then the cycles from $a$ can be pushed past $b$ if their elements do not occur in $b$.
   (c) $(5, 9)(1, 2, 6) \circ (1, 5, 6)(3, 9) \circ (6, 2, 1)(5, 9) = (1, 2, 9)(3, 5)$.

(d) Because of Theorem 4.24, one only needs to check how many permutations there are with which cycle structure: $24 = 1 + 3 + 6 + 6 + 8$.

## Section 7.4

1. $G$ cannot contain a glide reflection, because performing a glide reflection twice in a row yields a translation. If $G$ contains a rotation and another element that does not fix the rotation point, you can generate a translation from it. If $G$ does not contain a rotation, it can only contain a reflection.
2. Let $s$ be the reflection over a straight line through the origin. Then $sO_2^+$ is comprised of all reflections over straight lines through the origin.

## Section 7.5

1. Because $G$ operates flag-transitively, it operates transitively on the vertices, i.e., for any two vertices there is a group element that maps one vertex to the other. Therefore, any two vertices have the same degree. Since $G$ operates transitively on the $n$-gons, any two must have the same number of boundary edges. From flag-transitivity, it follows that any two interior angles and any two edges can be mapped onto each other, and thus all $n$-gons are regular $n$-gons.
2. $G_{(4,3)}^+ = \langle a, b, c \mid a^4, b^3, c^2, abc \rangle$.
3. You always have to tilt one of three mirrors backwards. For example, you can see two sides of a cube if you set up three mirrors to form an isosceles right triangle and tilt the mirror of one cathetus backwards by $45°$.

## Section 7.6

1. There are 12 edges and thus $2^{12}$ colorings. We have the same conjugation classes as in Anna's case. For each of the 5 conjugation classes, consider which cubes remain fixed. For example: For the 8 rotations around the cube diagonals, $2^4$ cubes remain fixed each time.
2. The group here is $D_8^+$, the group of rotations of the regular octagon. Because the group is abelian, each element is its own conjugation class. There are $2^8$ ways to place the candles, without considering rotations. If you carry out the calculation, you get

$$\frac{1}{8}(2^8 + 2 + 2^2 + 2 + 2^4 + 2 + 2^2 + 2) = 36$$

different ways.
3. Because $G$ operates transitively, there is only one orbit, and from Theorem 7.25 it follows that $|G| = \sum_{g \in G} |X^g|$. If $e \in G$ is the identity, then $|X^e| = |X| > 1$. If for all other $g \in G$ it holds that $|X^g| \geq 1$, then $\sum_{g \in G} |X^g| > |G|$.

## Section 8.1

1. If the generators of a group commute, then all group elements commute with each other.
2. $Q/Q' = \{Q', iQ', jQ', kQ'\}$ and $ij \in kQ'$, $ik \in jQ'$, $jk \in iQ'$, $i^2 \in Q'$, $j^2 \in Q'$, $k^2 \in Q'$, and thus $Q/Q'$ is isomorphic to the Klein four-group.
3. $xyx^{-1}y^{-1} = (yxy^{-1}x^{-1})^{-1}$.

4. The commutator subgroup is equal to $\mathcal{E}^+$. Commutators of reflections over parallel lines give all translations and commutators of reflections over lines which intersect give all rotations.

## Section 8.2

1. This follows directly from the classification Theorem 8.3.
2. Let $k$ be the least common multiple of $p$ and $q$. Then $|U| = k$ and $U \cong \mathbb{Z}_k$.

## Section 8.3

1. Let $G$ be solvable. Then, according to Lemma 8.9, there is a $k \in \mathbb{N}$ such that $G^{(k)} = \{e\}$. If $H$ is a subgroup of $G$, the assertion follows because $H^{(i)} < G^{(i)}$ (prove this using Lemma 8.15 for the homomorphism that embeds $H$ into $G$).
2. In Sect. 8.1, we saw that the commutator subgroup of $D_n$ is a group of rotations, which is abelian.
3. $H \lhd G$, $G/H = J$, and apply Theorem 8.16.
4. If $U$ is a subgroup of the nilpotent group $G$ with the central series $G = N_0 \rhd N_1 \rhd \cdots \rhd N_k = \{e\}$, then the groups $U \cap N_i$ form a central series for $U$.

## Section 9.2

1. Isometries preserve distances. If a straight line were mapped to a "curved" line by an isometry, lengths would shorten under the mapping.
2. It is to be proven that for any two lines $g, h$ in the hyperbolic plane, there is an isometry that maps $g$ onto $h$. To do this, map $g$ by a translation or a reflection so that it intersects with $h$. Then rotate around the intersection point.
3. No.

## Section 9.3

1. For example, if two vertices of the fundamental region lie on the boundary of the hyperbolic plane, and there is a $90°$ angle at the third vertex, we get the presentation $\langle a, b, c \mid a^2, b^2, c^2, (ab)^2 \rangle$.
2. (a) We obtain the Cayley graph by dualizing the decomposition. Around each vertex of the decomposition, we get a quadrilateral in the dual, which leads to relations of the type $(s_a s_b)^2$.
   (b) Each $d_i$ corresponds to a rotation by $180°$, and we get the relations $d_i^2$. Explain the last relation using Fig. 9.9.
3. $\langle s_1, s_2 \mid s_1^2, s_2^2, (s_1 s_2)^2 \rangle$.

## Section 10.1

1. Figure F.3 shows a van Kampen diagram for $d^2 s d^2 s$.
2. We eliminate the elements $-k, -i, -j$ and the relations $ji = -k, kj = -i, ik = -j$. We also eliminate $-1$ and $j^2 = -1$. Replace $i$ with $jk$ in the relations and obtain: $k^2 = j^2, k = jkj$ and $j = kjk$. Finally, show that the second relation is a consequence of the others.
3. First generate $a = ba^{-1}b$. That is $a^3 = ba^{-1}b^2a^{-1}b^2a^{-1}b$. It follows that $a^3 = ba^3b = b^4 = b^2a^3$ and therefore $b^2 = 1$ and thus $a^3 = 1$. These trivialization steps can be transformed into a van Kampen diagram.

**Fig. F.3**  van Kampen
Diagram for $d^2 s d^2 s$

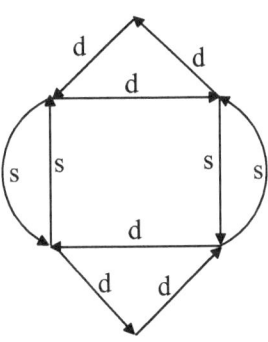

## Section 10.2

1. The center point $x$ of an edge in the Cayley graph $\Gamma$ is at a distance of $1/2$ from a boundary point $P$ of the edge. Therefore, $f(P)$ is at most $c/2$ away from $f(x)$. $g(f(x))$ then has at most distance $c \cdot c/2$ from $P$ and distance $(c^2 + 1)/2$ from $x$.

2. The mapping $f$ from the vertices of $\Gamma_\mathbb{Z}(1)$ to the vertices of $\Gamma_\mathbb{Z}(2,3)$ with $f(n) = n$ is a quasi-isometry with $\lambda = 3$ and $\epsilon = 0$.

## Section 10.3

1. Let $P$ be a given presentation and a van Kampen diagram for the relation $w$ be given. After multiplying a defining relation by another, the new van Kampen diagram for $w$ has at most twice as many disks as the old one.

2. We obtain the inequalities:

$$4S \leq 2K - k \tag{F.1}$$

and

$$5V - 3k \leq 2K - \bar{K}. \tag{F.2}$$

If we add four times (F.1) to three times (F.2), we obtain the following isoperimetric inequality after further transformations: $S \leq 5l(w) - 15$.

3. For example, the word $ba^{-1}$ appears exactly once in the defining relation or its inverse. The same applies to all other subwords of length 2 of the defining relation.

4.

```
gap> F:=FreeGroup(["j","k"]);;
gap> AssignGeneratorVariables(F);;
#I  Assigned the global variables [ j, k ]
gap> r1:=k^2*j^-2;; r2:=k*j*k*j^-1;;
gap> G:=F/[r1,r2];;
gap> PiecesOfGroup(G);
[ j^-1, j, k^-1, k, j^-1*k, j*k^-1, k^-1*j, k*j^-1 ]
gap> GroupSatisfiesC(G,4);
```

```
false
gap> GroupSatisfiesC(G,3);
false
gap> GroupSatisfiesC(G,2);
true
gap> GroupSatisfiesT(G,3);
true
gap> GroupSatisfiesT(G,4);
false
```

## Section 10.4

1. Yes. Their Cayley graphs are quasi-isometric to the Cayley graph $\Gamma_{\mathbb{Z}}(1)$.
2. The operation on the Cayley graph maps edges with their endpoints to edges with the corresponding images of the endpoints. It is also an isomorphism on the set of edges. Therefore, entire paths are mapped to paths of the same length. The operation is thus distance-preserving.
3. Draw the Cayley graph of this group, and follow the proof sketch of Example 10.25.

## Section 10.5

1. Let $|w_g| \leq c \cdot d_G(1, g)$ hold for all $g \in G$. Let $g \in G$, and the path $w_g$ consists of the points $1, g_1, g_2, \ldots, g_{n-1}, g$ (in that order). Let $\lambda, \epsilon$ be the constants from the definition of the quasi-isometry, and let $f(1) = 1$ in $\Gamma_H$. If we choose the path $w_{f(g)}$ appropriately from the images of the path of the preimage, it follows that

$$|w_{f(g)}| = d_H(f(1), f(g_1)) + d_H(f(g_1), f(g_2)) + \cdots + d_H(f(g_{n-1}), f(g))$$

$$= |w_g| \cdot (\lambda + \epsilon) \leq c \cdot d_G(1, g) \cdot (\lambda + \epsilon) \leq c' \cdot d_H(f(1), f(g)) \cdot (\lambda + \epsilon)$$

2. As with $\mathbb{Z} \times \mathbb{Z}$, sort the words by the generators.

# Appendix G
# Notes on the Literature

A beautiful introduction to Euclidean geometry and further reading on Chapter 1 can be found in [Jen97] and [AF09]. Elementary introductory books on group theory are: [Ale75, BN17, Far96, Fra89, Gal21, Glo16, Göt97, GM71, Mit77, Sie66]. The material of the first two chapters essentially comes from them. Parts of Chap. 3 and the section on the symmetric group in Chap. 4 also comes from these books. However, the presentation there is less geometric than here. Our approach is similar to [DC04]. In [Car09], many facts are explained particularly vividly and elementarily on the Cayley graph. Weyl [Wey55] provides a beautiful introduction to symmetry. See also [Ros75], Baldus and Rosebrock [BR98a] and Baldus and Rosebrock [BR98b] prepare the basics for school. Much of what is covered here can also be found in [Mei08] and is further developed there. The latter is a beautiful book, but it does not cover hyperbolic groups. Many of the geometric theorems in this book can be found in [Qua94].

Somewhat more demanding are [Art98], [Bud72], [Rob95], [KM09], [Ol'91] and [Sti96]. These books cover especially Chaps. 1–3 and deal with additional material from Chap. 4 (4.1–4.5), 5, 6 and 7. Sections 3.5 and 4.6 and further on group operations in geometry can be found in [CM79], very stimulating in [Lyn85], [Mag74], [Mil68] and [NST94]. Bridson and Haefliger [BH99] is relatively demanding. Thurston [Thu97] focuses more on geometry. In [Hen12], the friezes are classified. In [Ros05], initial approaches are presented as ideas for school.

Chapter 5 is further developed in [Coh89], [Joh90], [LS77] and [MKS76]. An overview of decision questions for group presentations can be found in [HR93a].

More detailed introductions to hyperbolic geometry than given in Chap. 9 can be found in the elementary and very beautiful [Sin97], [Fil93], [HCV96], [Ced91], [Cox63], [FG91], [NST94], and [Sti92]. Magnus [Mag74] goes into detail on decompositions of the hyperbolic plane. An easily readable introduction to Poincaré's circle model can be found in [Tru98] and in [AF09]. More details are in [Moi90].

© The Author(s), under exclusive license to Springer-Verlag GmbH, DE, part of Springer Nature 2024
S. Rosebrock, *Visual Group Theory*, Springer Undergraduate Mathematics Series, https://doi.org/10.1007/978-3-662-69365-0

The contents of Chap. 10 have only been lightly covered in books. You are most likely to find something in [Lö17, dlH00, CgRR08] but also in [Bau93], [BH99], [BRS07] or [CDP90]. There are many journal articles on this subject, see among many others [A⁺91, How99, HR93a, NS96] and the literature cited there. The best reference for Sect. 10.5 is [E⁺92].

Many books use matrix groups as examples of groups. More about matrix groups than given in the appendix can be found for example in [Arm88], [Art98], [Cig95], [NST94], [Rob95], and [Rot95].

# References

[A+91] J. Alonso et al., Notes on word hyperbolic groups, in *Group Theory from a Geometrical Viewpoint*, ed. by H. Short (World Scientific, Singapore, 1991), pp. 3–63

[AF09] I. Agricola, T. Friedrich, *Elementargeometrie* (Vieweg und teubner Verlag, Braunschweig, 2009)

[Ale75] P.S. Alexandroff, *Einführung in die Gruppentheorie*. VEB (Deutscher Verlag der Wissenschaften, Berlin, 1975). (Moskau 1951)

[Arm88] M.A. Armstrong, *Groups and Symmetry* (Springer, Berlin, 1988)

[Art98] M. Artin, *Algebra* (Birkhäuser Verlag, Basel, 1998)

[Bau93] G. Baumslag, *Topics in Combinatorial Group Theory*. Lectures in Mathematics, ETH Zürich (Birkhäuser Verlag, Basel, 1993)

[BH99] M. Bridson, A. Haefliger, *Metric Spaces of Non-Positive Curvature*. Grundlehren der mathematischen Wissenschaft, vol. 319 (Springer, Berlin, 1999)

[BN17] T. Barnard, H. Neill, *Discovering Group Theory* (CRC Press/Taylor and Francis, Boca Raton, 2017)

[Bog08] O. Bogopolski, *Introduction to Group Theory*. EMS Texbooks in Mathematics (European Mathematics Society Publishing House, Helsinki, 2008)

[BR98a] D. Baldus, S. Rosebrock, Isometrien und ihre Verkettungen (Teil I). Mathematik in der Schule **3**, 144–156 (1998)

[BR98b] D. Baldus, S. Rosebrock, Isometrien und ihre Verkettungen (Teil II). Mathematik in der Schule **4**, 209–220 (1998)

[BRS07] N. Brady, T. Riley, H. Short, *The Geometry of the Word Problem for Finitely Generated Groups*. Advanced Courses in Mathematics (Birkhäuser Verlag, Basel, 2007)

[Bud72] F.J. Budden, *The Fascination of Groups* (Cambridge University Press, Cambridge, 1972)

[Can02] J.W. Cannon, Geometric group theory, in *Handbook of Geometric Topology*, ed. by R.J. Daverman, R.B. Sher (Elsevier, Amsterdam, 2002), pp. 261–305

[Car09] N. Carter, *Visual Group Theory* (Mathematical Association of America, Washington, 2009)

[Car19] N. Carter, Group explorer 3.0 (2019). https://nathancarter.github.io/group-explorer/index.html

[CDP90] M. Coonaert, T. Delzant, A. Papadopoulos, *Géometrie et théorie des groupes, Les groupes hyperboliques de Gromov*. Lecture Notes in Mathematics, vol. 1441 (Springer, New York, 1990)

© The Author(s), under exclusive license to Springer-Verlag GmbH, DE, part of Springer Nature 2024
S. Rosebrock, *Visual Group Theory*, Springer Undergraduate Mathematics Series, https://doi.org/10.1007/978-3-662-69365-0

[Ce17] M. Clay, D. Margalit (eds.), *Office Hours with a Geometric Group Theorist* (Princeton University Press, Princeton, 2017)

[Ced91] J.N. Cederberg, *A Course in Modern Geometries*. Undergraduate Texts in Mathematics (Springer, New York, 1991)

[CgRR08] T. Camps, V.G. Rebel, G. Rosenberger, *Einführung in die kombinatorische und die geometrische Gruppentheorie*. Berliner Studienreihe zur Mathematik, Band 19 (Heldermann Verlag, Berlin, 2008)

[Cig95] J. Cigler, *Körper, Ringe, Gleichungen* (Spektrum Akademischer Verlag, Heidelberg, 1995)

[CM79] H.S.M. Coxeter, W.O.J. Moser, *Generators and Relations for Discrete Groups* (Springer, Berlin, 1979)

[Coh89] D.E. Cohen, *Combinatorial Group Theory: A Topological Approach*. London Mathematical Society Student Texts, vol. 14 (Cambridge University Press, Cambridge, 1989)

[Cos18] I.S. Costa, Small Cancellation – GAP Package (2018). http://mate.dm.uba.ar/~isadofschi

[Cox63] H.S.M. Coxeter, *Unvergängliche Geometrie* (Birkhäuser Verlag, Basel, 1963)

[DC04] S.V. Duzhin, B.D. Chebotarevsky, *Transformation Groups for Beginners*. Student Mathematical Library, vol. 25 (American Mathematical Society, Providence, 2004)

[dlH00] P. de la Harpe, *Topics in Geometric Group Theory*. Chicago Lectures in Mathematics (The University of Chicago Press, Chicago, 2000)

[E+92] D.B.A. Epstein, et al., *Word Processing in Groups* (Jones and Bartlett, Burlington, 1992)

[Far96] D.W. Farmer, *Groups and Symmetry*. Mathematical World, vol. 5 (American Mathematical Society, 1996)

[FG91] P.A. Firby, C.F. Gardiner, *Surface Topology* (Ellis Horwood Limited, 1991)

[Fil93] A. Filler, *Euklidische und nichteuklidische Geometrie* (BI Wissenschaftsverlag, Mannheim, 1993)

[Fis17] G. Fischer, *Lehrbuch der Algebra*. Springer Spektrum (Springer, New York, 2017)

[Fra89] J.B. Fraleigh, *A First Course in Abstract Algebra* (Addison Wesley, Boston, 1989)

[Gal21] J. Gallian, *Contemporary Abstract Algebra*. Textbooks in Mathematics, 10th edn. (Cengage Learning, Boston, 2021)

[GAP22] The GAP Group, *GAP – Groups, Algorithms, and Programming, Version 4.12.2* (2022)

[Glo16] T. Glosauer, *Elementar(st)e Gruppentheorie*. Springer Spektrum (Springer, New York, 2016)

[GM71] I. Grossmann, W. Magnus, *Gruppen und ihre Graphen* (Klett Studienbücher, Stuttgart, 1971)

[Göt97] P. Göthner, *Elemente der Algebra*. Mathematik-abc für das Lehramt. B.G.Teubner Verlagsgesellschaft (Stuttgart, Leipzig, 1997)

[Gro87] M. Gromov, Hyperbolic groups, in *Essays in Group Theory*, ed. by S. Gersten. M.S.R.I. Series, vol. 8 (Springer, Berlin, 1987), pp. 75–263

[HCV96] D. Hilbert, S. Cohn-Vossen, *Anschauliche Geometrie* (Springer, Berlin, 1996)

[Hen12] H.W. Henn, *Geometrie und Algebra im Wechselspiel*. Springer Spektrum (Springer, Berlin, 2012)

[How99] J. Howie, Hyperbolic groups lecture notes (Unpublished, 1999)

[HR93a] G. Huck, S. Rosebrock, Applications of diagrams to decision problems, in *Two-Dimensional Homotopy and Combinatorial Group Theory*, ed. by A. Sieradski C. Hog-Angeloni, W. Metzler. London Mathematical Society Lecture Note Series, vol. 197, pp. 189–218 (Cambridge University Press, Cambridge, 1993)

[HR93b] G. Huck, S. Rosebrock, A bicombing that implies a sub-exponential isoperimetric inequality. Proc. Edinburgh Math. Soc. **36**, 515–523 (1993)

[Jen97] G.A. Jennings, *Modern Geometry with Applications*. Universitext (Springer, New York, 1997)

[Joh90] D.L. Johnson, *Presentations of Groups*. London Mathematical Society Student Texts, vol. 15 (Cambridge University Press, Cambridge, 1990)

[Joy02] D. Joyner, *Adventures in Group Theory* (The Johns Hopkins University Press, Baltimore, 2002)

[KM09] C. Karpfinger, K. Meyerberg, *Algebra* (Spektrum Akademischer Verlag, Heidelberg, 2009)

[LS77] R. Lyndon, P. Schupp, *Combinatorial Group Theory* (Springer, Berlin, 1977)

[Lyn85] R.C. Lyndon, *Groups and Geometry*. London Mathematical Society Lecture Note Series, vol. 101 (Cambridge University Press, Cambridge, 1985)

[Lö17] C. Löh, *Geometric Group Theory*. Universitext (Springer, New York, 2017)

[Mag74] W. Magnus, *Non-euclidian Tesselations and Their Groups* (Academic Press, New York, 1974)

[Mei08] J. Meier, *Groups, Graphs and Trees*. London Mathematical Society Student Texts, vol. 73 (Cambridge University Press, Cambridge, 2008)

[Mil68] J. Milnor, A note on curvature and fundamental group. J. Differ. Geom. **2**, 1–7 (1968)

[Mit77] A. Mitschka, *Elemente der Gruppentheorie* (Herder Verlag, Freiburg im Breisgau, 1977)

[MKS76] W. Magnus, A. Karrass, D. Solitar, *Combinatorial Group Theory* (Dover Publications, New York, 1976)

[Moi90] E.E. Moise, *Elementary Geometry from an Advanced Standpoint*, 3rd edn. (Addison-Wesley, Boston, 1990)

[NS96] W. Neumann, M. Shapiro, A short course in geometric group theory, in *Notes for the ANU Workshop* (1996)

[NST94] P. Neumann, G. Stoy, E. Thompson, *Groups and Geometry* (Oxford Science Publications, Oxford, 1994)

[Ol'91] A.Y. Ol'shanskii, *Geometry of Defining Relations in Groups*. Mathematics and its Applications (Kluwer Academic Publishers, Dordrecht, 1991)

[Qua94] E. Quaisser, *Diskrete Geometrie* (Spektrum Akademischer Verlag, Heidelberg, 1994)

[Rat94] J.G. Ratcliffe, *Foundations of Hyperbolic Manifolds*. Graduate Texts in Mathematics, vol. 149 (Springer Verlag, New York, 1994)

[Rob95] D.J.S. Robinson, *A Course in the Theory of Groups*. Graduate Texts in Mathematics, vol. 80 (Springer, New York, 1995)

[Ros75] J. Rosen, *Symmetry Discovered* (Cambridge University Press, Cambridge, 1975)

[Ros05] S. Rosebrock, Aus Spiegelachsen Figuren bauen. Mathematikinformation **42**, 59–65 (2005)

[Rot95] J.J. Rotman, *An Introduction to the Theory of Groups*. Graduate Texts in Mathematics, vol. 148 (Springer, New York, 1995)

[Sie66] K. Sielaff, *Einführung in die Theorie der Gruppen*. Schriftenreihe zur Mathematik, vol. 4 (Otto Salle Verlag, Frankfurt, 1966)

[Sin97] D.A. Singer, *Geometry: Plane and Fancy*. Undergraduate Texts in Mathematics (Springer, New York, 1997)

[Sti92] J. Stillwell, *Geometry of Surfaces*. Universitext (Springer, New York, 1992)

[Sti96] J. Stillwell, *Elements of Algebra*. Undergraduate Texts in Mathematics (Springer, New York, 1996)

[Thu97] W. Thurston, *Three-dimensional Geometry and Topology, Volume I*. Princeton Mathematical Series (Princeton University Press, Princeton, 1997)

[Tru98] R. Trudeau, *Die geometrische Revolution* (Birkhäuser Verlag, Basel, 1998)

[Wey55] H. Weyl, *Symmetrie* (Birkhäuser Verlag, Basel, 1955)

# Index